Charles Edward White

The senior arithmetic for grammar schools

Charles Edward White

The senior arithmetic for grammar schools

ISBN/EAN: 9783337274887

Printed in Europe, USA, Canada, Australia, Japan

Cover: Foto ©Paul-Georg Meister /pixelio.de

More available books at **www.hansebooks.com**

THE
SENIOR ARITHMETIC

FOR

GRAMMAR SCHOOLS

BY

CHARLES E. WHITE

PRINCIPAL FRANKLIN SCHOOL, SYRACUSE, N.Y.

BOSTON, U.S.A.
D. C. HEATH & CO., PUBLISHERS
1897

PREFACE.

The Senior Arithmetic is intended for use in the higher grammar grades, beginning with the sixth year.

The first eight pages are devoted to definitions in review, covering fundamental subjects which should be mastered before the book is taken up. The senior arithmetic proper begins with decimals. Though largely review matter, denominate numbers are treated with considerable fulness. Teachers are expected, however, to use their discretion as to omissions in this subject.

It has been the author's aim to employ such definitions, solutions, explanations, and rules as can be readily comprehended and applied by the pupils, unaided by the teacher. While care has been exercised in selecting a great variety of practical business problems and in arranging them progressively, the development of mental power has been kept constantly in view.

The arrangement of this book is topical, but subjects previously studied are kept fresh in the minds of the pupils by frequent carefully prepared reviews.

The practice of referring percentage problems back to the original questions of relation has proven highly successful in the author's experience.

Thanks are due to the various superintendents of city schools who kindly furnished copies of recent examination questions, which largely constitute the test problems of this book.

The author has also received invaluable aid from many leading educators in the State of New York, all of whom he desires to thank most cordially.

C. E. W.

SYRACUSE, N.Y., *Oct.* 18, 1896.

CONTENTS.

DEFINITIONS, RULES, AND PRINCIPLES IN REVIEW —   PAGE
  Notation, Numeration 1
  Addition, Subtraction, Multiplication 2
  Division, Factors, Multiples 3
  Least Common Multiple 4
  Greatest Common Divisor 4
  Cancellation 5
  Common Fractions 5
  Addition and Subtraction of Fractions 7
  Multiplication and Division of Fractions 8

DECIMAL FRACTIONS —
  To read a decimal 9
  To write a decimal 10
  To reduce two or more decimals to a common denominator . . 11
  To reduce a common fraction to a decimal 12
  Addition of decimals 13
  Subtraction of decimals 13
  Multiplication of decimals 14
    To multiply by 10, 100, 1000, etc. 15
  Division of decimals 16
    To divide by 10, 100, 1000, etc. 17
  Parts of 100 and 1000 18
  To multiply by 25 18
  Aliquot Parts of a dollar 19
  Review of decimals 21
  Accounts and Bills 25
  Indicated Operations 27

MISCELLANEOUS REVIEW —
  Factors, Multiples, Divisors, and Cancellation 30
  Common Fractions 32
  Review questions 39

COMPOUND NUMBERS —
  Linear measure 43
  Surveyors' measure 44

CONTENTS.

COMPOUND NUMBERS (*Continued*) —   PAGE
   Square measure 44
   Cubic measure 44
   Liquid measure 45
   Apothecaries' fluid measure 45
   Dry measure 45
   Avoirdupois weight 45
   Troy weight 46
   Apothecaries' weight 46
   Measure of time 46
   Circular measure 47
   Federal money 48
   English or Sterling money 48
   Counting table 49
   Paper table 49
   Reduction descending 49
   Reduction ascending 51
   Review problems 53
   To reduce denominate fractions to integers of lower denominations 55
   To reduce denominate numbers to fractions of higher denominations 56
   To find what part one denominate number is of another . . . 58
   Addition of compound numbers 58
   Subtraction of compound numbers 61
   Difference between dates 63
   Multiplication of compound numbers 64
   Division of compound numbers 65
   Miscellaneous problems 69

MEASUREMENTS, Surfaces 72
   Carpeting rooms 77
   Plastering and painting 79
   Papering walls 81
   Board measure 82
   Miscellaneous problems 84

MEASUREMENTS, Solids 86
   Wood measure 89
   Capacity of bins 90

LONGITUDE AND TIME 91
   Standard time 92
   Review questions 94

THE METRIC SYSTEM —
   Linear measure 96
   Surface measure 100

CONTENTS.

The Metric System (*Continued*) —   PAGE
  Volume measure 102
  Capacity measure 104
  Measure of weight 105
  Review questions 106
  General review 107

Percentage 114
  Profit and loss 124
  Commission 126
  Insurance 130
  Trade discount 132
  Taxes 134
  Duties 136
  Review questions 137
  Miscellaneous review of percentage 137

Simple Interest 143
  Exact interest 149
  To find the rate, when principal, interest, and time are given . 150
  To find time, when principal, interest, and rate are given . . 151
  To find principal, when interest or amount, rate and time are
    given 152
  Promissory notes 153
  Partial payments, U. S. rule 156
  Merchants' rule 159
  Compound interest 160
  Review of interest 161

True Discount 165

Bank Discount 167
  To find the face of a note when the proceeds, time, and rate are
    known 173
  Review of discount 174

Stocks and Bonds 176
  Bonds 179
  Miscellaneous 182

Average of Payments 183
  Review questions 188

Ratio and Proportion —
  Ratio 189
  Simple proportion 190
  Compound proportion 195
  Partnership 198
  Review questions 202

CONTENTS.

INVOLUTION AND EVOLUTION —   PAGE
- Involution 203
- Evolution 204
- Square root 205
- Right-angled triangles 211
- Similar surfaces 213
- Cube root 214
- Similar solids 220
- Questions 221
- General review 222

TEST QUESTIONS 236
- Decimals 244
- Denominate numbers 248
- Percentage 255
- Interest and discount 263
- Proportion and partnership 271
- Involution and evolution 274
- Miscellaneous 276

MENSURATION —
- Surfaces 285
- Solids 287
- Review of mensuration 290

SENIOR ARITHMETIC.

DEFINITIONS.

1. A **Unit** is one, or one thing.

2. A **Number** is that which tells how many.

3. The **Unit of a Number** is one of its units.

4. Numbers having the same unit are **Like Numbers.**

5. A number not applied to any particular object is an **Abstract Number**; as 6, 11, 15.

6. A number that is applied to a particular object is a **Concrete Number**; as 6 men, 11 lb., 15 days.

7. An **Integer** is a whole number.

8. Expressing numbers by figures or letters is called **Notation.**

9. **Arabic Notation** is expressing numbers by figures.

10. **Roman Notation** is expressing numbers by letters.

11. Naming the places of figures and reading numbers is **Numeration.**

12. A figure standing alone expresses units.

13. When figures stand side by side, the right-hand figure expresses units, the next tens, the next hundreds, etc.

14. The value of a figure, without regard to its place, is its **Simple Value.** The value of a figure with reference to its place in a number is its **Local Value.**

NOTE.— In the number 5555, the simple value of each figure is 5. The local value of the right-hand figure is 5. Of the second, 50. Of the third, 500. Of the fourth, 5000.

15. Each group of three figures, beginning with units and counting to the left, is a **Period.**

TO READ NUMBERS.

Rule. — Begin at the right, and separate the numbers into groups of three figures each, using the comma.

Begin at the left, and read the number in each group, giving to it the name of that group.

No name is given to the number in the last group.

16. Addition is the process of uniting two or more like numbers into one sum.

17. The result of addition is called the **Sum** or **Amount.**

18. Subtraction is the process of finding the difference between two like numbers.

19. The number from which we subtract is called the **Minuend,** and the number subtracted, the **Subtrahend.** The result in subtraction is the **Difference** or **Remainder.**

20. Multiplication is the process of finding a number that is a given number of times another number.

21. The **Multiplicand** is the number multiplied.

22. The **Multiplier** is the number multiplied by.

23. The result of multiplication is called the **Product.**

PRINCIPLES.— The multiplier must be an abstract number. The multiplicand and product are like numbers. The product is the same in whatever order the numbers are taken.

DEFINITIONS.

24. Division is the process of finding how many times one number is contained in another.

25. The number divided is the **Dividend.**

26. The number by which the dividend is divided is the **Divisor.**

27. The result of division is the **Quotient.**

28. When the divisor is not exactly contained in the dividend, the part of the dividend that is left is the **Remainder.**

29. PRINCIPLES. — The remainder and dividend are like numbers. When the divisor is abstract, the dividend and quotient are like numbers. When the dividend and divisor are concrete, the quotient is abstract.

30. The **Sign of Division** is ÷, and when placed between two numbers signifies that the first is to be divided by the second.

FACTORS AND MULTIPLES.

31. A **Factor** of a number is any integer that will exactly divide it.

32. A number that has no factors except itself and 1 is a **Prime Number.**

33. A number that has other factors besides itself and 1 is a **Composite Number.**

34. A prime number used as a factor is a **Prime Factor.**

35. What are the prime factors of 1155?

Rule.— *Divide the number by any prime factor that will exactly divide it. Divide the quotient in the same manner. Continue the division until a quotient is found that is a prime number.*

5 / 1155
3 / 231
7 / 77
11

5, 3, 7, and 11 *Ans.*

The divisors and the last quotient are the prime factors.

36. Numbers that have no common factor or divisor are **Prime to Each Other.**

37. A **Multiple** of a number is a number that exactly contains that number. 15 is a multiple of 5.

38. A number that is a multiple of two or more numbers is a **Common Multiple** of them. 24 is a common multiple of 4 and 3.

39. The least multiple of two or more numbers is their **Least Common Multiple.** 12 is the least common multiple of 3 and 4.

40. What is the least common multiple of 18, 27, and 30?

Rule. — *Divide by any prime number that is exactly contained in two or more of the numbers, and bring down the quotient and undivided numbers. Divide again as before, continuing the division until the quotients and undivided numbers are prime to each other.*

The product of the divisors, quotients, and undivided numbers is the least common multiple.

$$2 \underline{)\, 18,\ 27,\ 30}$$
$$3 \underline{)\, 9,\ 27,\ 15}$$
$$3 \underline{)\, 3,\ 9,\ 5}$$
$$\ 1,\ 3,\ 5$$

$2 \times 3 \times 3 \times 3 \times 5 = 270.$ *Ans.*

41. A number that is a factor of two or more numbers is a **Common Divisor** of them. 5 is a common divisor of 30 and 40.

42. The greatest factor of two or more numbers is the **Greatest Common Divisor** of them. 10 is the greatest common divisor of 30 and 40.

43. What is the greatest common divisor of 324 and 372?

Rule. — Divide the greater number by the less, then the divisor by the remainder, until there is no remainder. The last divisor is the greatest common divisor.

```
324 / 372 ( 1
324
 48 / 324 ( 6
      288
       36 / 48 ( 1
            36
            12 / 36 ( 3
```

When there are more than two numbers, first find the greatest common divisor of two of them, then of this divisor, and a third number, until all the numbers are used.

CANCELLATION.

44. Cancellation is a process of shortening indicated division by rejecting the same factors from both dividend and divisor.

45. PRINCIPLES. — Rejecting the same factor from dividend and divisor divides both by that factor.

Dividing both dividend and divisor by the same number does not affect the quotient.

COMMON FRACTIONS.

46. A **Fraction** is one or more of the equal parts of a unit. The unit of which the fraction is a part is called the **Unit of the Fraction**, and one of the equal parts is called the **Fractional Unit**. Two or more fractions having the same fractional unit are **Like Fractions**.

47. A fraction is written with two numbers, one above the other, with a line between; as, $\frac{2}{3}$.

48. The number below the line in a fraction is the **Denominator**, and shows into how many equal parts the unit is divided.

49. The number above the line in a fraction is the **Numerator**, and shows how many of the parts are taken.

50. The numerator and denominator are the **Terms of a Fraction**.

51. A **Proper Fraction** is a fraction whose value is less than 1. Its numerator is less than its denominator; as, $\tfrac{3}{4}, \tfrac{1}{2}$.

52. An **Improper Fraction** is a fraction whose value is 1 or more than 1. Its numerator is equal to, or greater than, its denominator; as, $\tfrac{4}{4}, \tfrac{5}{4}, \tfrac{36}{11}$.

53. An integer may be written in fractional form by giving it 1 for a denominator; as, $5 = \tfrac{5}{1}$.

54. A **Mixed Number** is a number composed of an integer and a fraction; as, $3\tfrac{1}{4}, 5\tfrac{7}{8}$.

55. Reduction of Fractions is the process of changing their forms without changing their values.

56. PRINCIPLE. — Multiplying or dividing both terms of a fraction by the same number does not change the value of the fraction.

57. To reduce a fraction to Higher Terms.

Rule. — *Multiply both terms by the same number.*

58. $\tfrac{3}{5}$ equals how many 60ths?

NOTE. — Since the new denominator must be 60, or twelve times the given denominator, the new numerator must be twelve times the given numerator, therefore $\tfrac{3}{5} = \tfrac{18}{60}$.

59. To reduce a fraction to its Lowest Terms.

Rule. — *Divide both terms by any common factor; divide the result in the same way until the terms are prime to each other. If the terms are large, divide by their greatest common divisor.*

60. To reduce a Mixed Number to an Improper Fraction.

Rule. — *Multiply the integer by the denominator of the fraction, add the numerator to the product, and write the result over the denominator.*

DEFINITIONS.

61. To reduce an Improper Fraction to an Integer or Mixed Number.

Rule. — Divide the numerator by the denominator.

62. A number that is the denominator of two or more fractions is the **Common Denominator** of those fractions.

63. NOTE 1. — The common denominator of two or more fractions is a common multiple of their denominators.

NOTE 2. — The least common denominator of two or more fractions is the least common multiple of their denominators.

64. To reduce fractions to a Common Denominator.

Rule. — Multiply the denominators together for the common denominator, divide it by the denominator of each fraction, and multiply both terms by the quotient.

NOTE. — To find the least common denominator, find the least common multiple of the denominators, and proceed as above.

ADDITION OF FRACTIONS.

65. *Rule. — If the fractions are not like fractions, reduce them to a common denominator, add their numerators, and place the sum over the common denominator. Reduce the result to lowest terms. If the result is an improper fraction, reduce it to an integer or mixed number.*

To add mixed numbers, add the integers and fractions separately, and unite the results.

SUBTRACTION OF FRACTIONS.

66. *Rule. — If the fractions are not like fractions, reduce them to a common denominator, and write the difference of their numerators over the common denominator.*

To subtract mixed numbers, subtract integers and fractions separately.

MULTIPLICATION OF FRACTIONS.

67. Rule.—*Reduce integers and mixed numbers to improper fractions, and multiply the numerators together for the numerator of the product, and the denominators for the denominator of the product.*

Cancel when possible.

68. Two or more fractions joined by "of" form a **Compound Fraction.** The word "of" is equivalent to the sign of multiplication.

DIVISION OF FRACTIONS.

69. Principle. — To divide by a fraction is to multiply by that fraction inverted.

70. Rule.—*Reduce integers and mixed numbers to improper fractions, and multiply the dividend by the divisor inverted.*

Cancel when possible.

71. $\frac{7}{8} \div \frac{3}{4}$ may be written $\frac{\frac{7}{8}}{\frac{3}{4}}$. Such an expression is called a **Complex Fraction,** and is used simply to indicate division.

72. A **Complex Fraction** is a fraction having a fraction in one or both of its terms.

DECIMAL FRACTIONS.

73. A **Power** is the product of equal factors, as $5 \times 5 = 25$, $5 \times 5 \times 5 = 125$. 25 is the second power of 5. 125 is the third power of 5. $10 \times 10 = 100$. $10 \times 10 \times 10 = 1000$. 100 is the second power of 10. 1000 is the third power of 10.

74. A **Decimal Fraction** or **Decimal** is a fraction whose denominator is 10 or a power of 10.

DECIMAL FRACTIONS.

NOTE. — The denominator of a common fraction may be any number, but the denominator of a decimal fraction must be 10, 100, or 1000, etc.

75. A decimal is written at the right of a period (.) called the **Decimal Point**.

NOTE. — It is not customary to write the denominator of a decimal. It is determined by the position of the decimal point.

76. A figure at the right of a decimal point is called a **Decimal Figure**. Tenths are written like dimes, with one decimal figure. Thus, $\frac{5}{10} = .5$. Hundredths are written like cents, with two decimal figures.

Thus, $\frac{25}{100} = .25$; $\frac{7}{100} = .07$.

Thousandths are written like mills, with three decimal figures; thus, $\frac{125}{1000} = .125$; $\frac{16}{1000} = .016$; $\frac{4}{1000} = .004$. Ten-thousandths require four decimal figures; hundred-thousandths, five; millionths, six, etc.

77. Name the denominators in the following: .36; .08; .294; .1406; .0001; .263402.

Change to decimals: $\frac{25}{100}$; $\frac{125}{1000}$; $\frac{1063}{100000}$; $\frac{36}{1000000}$; $\frac{5}{1000}$; $\frac{1}{10000}$.

78. A **Mixed Decimal** is an integer and a decimal; as, 16.04.

79. To read a decimal.

Rule. — Read the decimal as an integer, and give it the denomination of the right-hand figure.

Read the following numbers:

1. .7
2. .07
3. .007
4. .700
5. .03065
6. .16984
7. .10016
8. .0000054
9. 35.18006
10. .0005
11. .500
12. 4.98625
13. 38694.06
14. 9.98463004
15. 235.850062

16. 100.000104   18. 3543.4536982   20. 303.303303
17. 9.1632002   19. 30.3303303   21. 9.999999

80. To write a decimal.

Rule.— Write the numerator, prefixing ciphers when necessary to express the denominator, and place the point at the left.

NOTE. — There must be as many decimal places in the decimal as there are ciphers in the denominator.

Express decimally :

22. Four tenths. Seventeen hundredths. Five hundredths. Three hundred twenty-five thousandths. Five thousandths. Fifteen thousandths. Nineteen, and seven hundred twenty-four thousandths.

23. Seven thousand five hundred four ten-thousandths. Sixteen, and 125 ten-thousandths. Six ten-thousandths. Five thousand ten-thousandths.

24. Seventeen thousand two hundred eleven hundred-thousandths. Four hundred-thousandths. Fifteen hundred-thousandths. Eighteen, and two hundred sixteen hundred-thousandths. One hundred twelve hundred-thousandths.

25. Twenty-nine hundredths. Twenty-nine thousandths. Twenty-nine ten-thousandths. Twenty-nine hundred-thousandths. One and one tenth. One and one hundredth. One and one thousandth. One and one ten-thousandth. One and one hundred-thousandth.

26. 324 and one hundred twenty-six millionths. 4582 and 36242 hundred-thousandths. Seventeen millionths. Five hundred-thousandths. Twenty-four, and three thousand four hundred six ten-millionths.

27. 10 millionths. 824 ten-thousandths. 31 hundredths.

216 hundred-thousandths. 7846 hundred-millionths. Four and 15 hundred-thousandths.

28. $\frac{8}{10}$
29. $\frac{15}{100}$
30. $\frac{615}{1000}$
31. $\frac{2123}{10000}$
32. $\frac{289}{100000}$
33. $\frac{28854}{1000000}$
34. $\frac{563}{10000000}$
35. $15\frac{5}{1000}$
36. $\frac{3}{10}$
37. $\frac{1}{100}$
38. $500\frac{5}{10}$
39. $\frac{27}{10000}$
40. $\frac{1}{1000000}$
41. $\frac{1000}{1000}$
42. $\frac{5}{100000}$
43. $\frac{275}{10000}$

REDUCTION OF DECIMALS.

81. PRINCIPLES. — Ciphers annexed to decimals do not change their value.

For each cipher prefixed to a decimal, the value is diminished ten-fold.

The denominator of a decimal when expressed is always 1 with as many ciphers as there are decimal places in the decimal.

82. To reduce two or more decimals to a Common Denominator.

Rule. — *Annex ciphers so that each decimal will have the same number of decimal figures.*

83. Reduce to a common denominator:

44. .5, .017, .1256, .000155, 29.803.
45. .80062, 305.24, 70.5, 3.85263.
46. .1, .0001, 1000.001, 1 .0100385.
47. .26, .13682, 9.4, 25., 8.63521.

84. Reduce .375 to a common fraction.

.375 as a common fraction is $\frac{375}{1000}$. This in lowest terms = $\frac{3}{8}$.

Rule. — *Write the numerator, omitting the point. Supply the denominator, and reduce to lowest terms.*

Reduce to common fractions:

48.	1.24	53.	.325	58.	16.144
49.	.16	54.	.113	59.	28.3695
50.	.325	55.	.7282	60.	34.000010
51.	.098	56.	2.25	61.	25.0000100
52.	.875	57.	.2425	62.	1084.0025

85. 63. Reduce $.37\tfrac{1}{2}$ to a common fraction.

Solution. — $\dfrac{37\tfrac{1}{2}}{100} = \dfrac{\tfrac{75}{2}}{100} = \dfrac{75}{200} = \tfrac{3}{8}$. *Ans.*

64.	$.12\tfrac{1}{2}$	67.	$.16\tfrac{2}{3}$	70.	$.87\tfrac{1}{2}$
65.	$.06\tfrac{1}{4}$	68.	$.33\tfrac{1}{3}$	71.	$.66\tfrac{2}{3}$
66.	$.62\tfrac{1}{2}$	69.	$.83\tfrac{1}{3}$	72.	$.36\tfrac{7}{8}$

86. To reduce a common fraction to a Decimal.

Reduce $\tfrac{3}{4}$ to a decimal.

$\tfrac{3}{4} = 3$ times $\tfrac{1}{4}$. $3 = (3.0)$, 30 tenths. $\tfrac{1}{4}$ of $3.0 = (.7)$, 7 tenths, and 2 tenths remainder. 2 tenths = 20 hundredths. $\tfrac{1}{4}$ of $.20 = .05$. Hence $\tfrac{3}{4} = .7 + .05 = .75$.

Rule. — *Annex decimal ciphers to the numerator, and divide by the denominator. Point off from the right of the quotient as many places as there are ciphers annexed.*

Notes. — A decimal cipher is a cipher at the right of the decimal point. If there are not enough figures in the quotient, prefix ciphers. The division will not always be exact. In such cases write the remainder over the divisor as a common fraction, or place the sign + after the decimal to show that the result is incomplete. Thus, $\tfrac{1}{7} = .142\tfrac{6}{7}$ or $.142 +$.

87. Reduce to decimals:

73.	$\tfrac{4}{5}$	77.	$\tfrac{3}{16}$	81.	$\tfrac{5}{8}$	85.	$\tfrac{7}{16}$	89.	$66\tfrac{2}{3}$
74.	$\tfrac{5}{8}$	78.	$\tfrac{5}{9}$	82.	$\tfrac{7}{24}$	86.	$\tfrac{18}{20}$	90.	$25.12\tfrac{1}{2}$
75.	$\tfrac{3}{4}$	79.	$\tfrac{3}{7}$	83.	$\tfrac{7}{8}$	87.	$12\tfrac{1}{2}$	91.	$16\tfrac{1}{4}$
76.	$\tfrac{2}{3}$	80.	$\tfrac{1}{2}$	84.	$\tfrac{6}{11}$	88.	$33\tfrac{1}{3}$	92.	$16.25\tfrac{1}{8}$

DECIMAL FRACTIONS.

ADDITION.

88. Add .35, 4.375, 28.3065.

Rule. — *Write the numbers so that decimal points stand in a column. Add as in integers, and place the point in the sum directly under the points above.*

$$\begin{array}{r} .35 \\ 4.375 \\ 28.3065 \\ \hline 33.0315 \end{array}$$

Find the sum:

93.	24.36	94.	38.28006	95.	1.186
	1.358		1.005		.285
	.004		2.16		.003
	1632.1		1873.148½		203.

96. .175 + 1.75 + 17.5 + 175. + 1750.

97. 145. + 14.5 + 1.45 + .145 + .0145.

98. 32.58 + 28963.1 + 287.531 + 76398.9341.

99. 1. + .1 + .01 + .001 + 100 + 10. + 10.1 + 100.001.

100. 1.923 + .008 + 251.47 + 1.961 + 0.0543 + .006 + 18.7.

101. Add 750.3521, 698.42001, .005321, 3.5, 749.006984, 36950.06, 875.942, 286.753.

102. Add 5 tenths; 8063 millionths; 25 hundred-thousandths; 48 thousandths; 17 millionths; 95 ten-millionths; 5, and 5 hundred-thousandths; 17 ten-thousandths.

103. Add $24\frac{3}{4}$, $17\frac{1}{4}$, .0058, $7\frac{1}{8}$, $9\frac{1}{16}$.

SUBTRACTION.

89. *Rule.* — *Write the numbers so that the decimal point of the subtrahend stands directly under the decimal point in the minuend. Subtract as in integers, and place the point directly under the points above.*

NOTE. — It is sometimes convenient to give the decimals the same denominator by annexing ciphers.

104. From 6.008   105. 38.   106. 26.34   107. 16.2600
    Take 3.154   .356   1.28983   1.0001

108. 32.90596 − 75
109. 9.5 − 3.35006
110. 856.2 − 8.562
111. .1 − .00001
112. 1000 − .001
113. 20 − .00205

114. .00011 − .000011
115. 10 − .1 + .0001
116. 8.75 + .95 + .125
117. 16 − .00001 + 27.69852
118. 2.5 − .09 + 1.85 − 1.283
119. 83.1 − 8.31 + .831

120. From one thousand take five thousandths.

121. Take 17 hundred-thousandths from 1.2.

122. From 8.5 take eighty-four hundredths.

123. Find the sum of 500 thousandths and 5 hundred-thousandths and from it subtract $\frac{3}{10}$.

124. From $17.37\frac{1}{2}$ take $14.16\frac{1}{8}$.

125. Find the difference between $\frac{384}{1000}$ and $\frac{384}{10000}$.

126. From 10 take $\frac{1}{10}$; $\frac{1}{100}$; 4.98; 1.05.

127. From one million and one millionth take one tenth.

128. From 1 tenth take 1 millionth.

129. Which is the greater and how much, one tenth or 100 thousandths?

130. Prove that $\frac{1}{2}$ and .500 are equal.

MULTIPLICATION.

90. Every decimal equals a corresponding common fraction, and for each cipher in its denominator there is a decimal figure in the decimal fraction.

$\frac{5}{100} \times \frac{3}{10} = \frac{15}{1000}$. (Three ciphers in the denominator.)
.05 × .3 = .015. (Three decimal places in the decimal.)

Rule. — *Multiply as in integers, and give to the product as many decimal figures as there are in both multiplier and multiplicand.*

DECIMAL FRACTIONS. 15

NOTE. — If there are not figures enough, prefix ciphers.
Ciphers at the right of a decimal have no value, and may be omitted.

Find the products:

1. .38 × 1.6.
2. .015 × .05.
3. 7½ × 3.4.
4. 50 × .304.
5. 2.65 × .104.
6. 257 × .354.
7. .296 × 124.
8. 1.001 × 1.01.
9. 13.33 × 1.3.
10. 25.863 × 4¼.

11. 1.04 × 6½.
12. 327⅜ × 4⅔.
13. 58.42 × 20.06.
14. .0001 × 1000.
15. .325 × 12½.
16. .333 × .333.
17. .001542 × .0052.
18. 26 × 36.82.
19. 2.84 × 3¼.
20. 11.11 × 100.

91. To multiply by 10, 100, 1000, etc.

21. Multiply 1.265 by 100.

Remove the point one place to the right for each cipher in the multiplier.
Do not write the multiplier.

$$\begin{array}{r} 1.265 \\ 100 \\ \hline 126.500 \end{array}$$

Oral.

22. 3689.25 × 10.
23. 38.6422 × 100.
24. 269.8342 × 1000.
25. 100 × 23.85.
26. 1000 × 1.52.

27. .5 × 100.
28. .5 × 1000.
29. 384.2 × 10.
30. .3659 × 100.
31. .1000 × .01.

92. To multiply by 200, remove the point to the right and multiply by 2.

Oral.

32. 86.44 × 200.
33. 3.894 × 3000.
34. 88.42 × 20.

35. 750.5 × 5000.
36. 1.892 × 2000.
37. 156.2 × 200.

93. Written.

38. Find the product of 1 thousand by one thousandth. 1 million by one millionth.

39. Multiply 700 thousandths by 7 hundred-thousandths.

40. Multiply the sum of 2 millionths and 10 thousandths by their difference.

41. Multiply together .35, 18.5, 28.004.

DIVISION.

94. Since in multiplication there are as many decimal places in the product as in both multiplier and multiplicand, in division the quotient must have as many places as the number of places in the dividend exceeds those in the divisor.

1. Divide 12.685 by .5.

SOLUTION. — Since there are three decimal places in the dividend and one in the divisor, there must be two in the quotient.

$$.5\,\overline{)\,12.685} \atop 25.37$$

Rule I. — *In all cases divide as in integers, then place the decimal point.*

2. Divide 399.552 by 192.

Rule II. — *When the divisor is an integer, place the point in the quotient directly over the point in the dividend in long division (directly under in short division). Prove by multiplying divisor by quotient.*

$$\begin{array}{r} 2.081 \\ 192\,\overline{)\,399.552} \\ 384 \\ \hline 1555 \\ 1536 \\ \hline 192 \\ 192 \end{array}$$

PRINCIPLE. — Multiplying both dividend and divisor by the same number does not change the quotient.

3. Divide 28.78884 by 1.25.

DECIMAL FRACTIONS.

Rule III. — *When the divisor contains decimal figures, move the point in both divisor and dividend as many places to the right as there are decimal places in the divisor (this, in Ex. 3, multiplies both by 100), then place the point in the quotient as if the divisor were an integer.*

```
                    23.031+
        1.25 / 28.78.884
                 250
                 ---
                 378
                 375
                 ---
                 388
                 375
                 ---
                 134
                 125
                 ---
                   9
```

NOTE 1. — The new points may be placed on a line with the tops of the figures, and the original point may stand to preserve the reading of the decimals.

NOTE 2. — If the quotient does not have a sufficient number of figures, prefix ciphers.

NOTE 3. — Before commencing to divide, see that there are at least as many decimal places in the dividend as in the divisor.

NOTE 4. — If there is a remainder after all the figures of the dividend are used, annex decimal ciphers and continue the division.

NOTE 5. — It is not usually necessary to have more than four decimal figures in the quotient.

Find the quotients:

1. .288 ÷ .64.
2. .36 ÷ 600.
3. 144 ÷ .12.
4. .25 ÷ .2500.
5. .12 ÷ 30.
6. .96 ÷ .08.
7. 384.526 ÷ 1.16.
8. 1440 ÷ .0018.
9. 1.225 ÷ 4.9.
10. 9.156 ÷ 12.
11. 315.432 ÷ .132.
12. 1.5906 ÷ 241.
13. 36.25 ÷ 1.25.
14. 75 ÷ .0125.
15. 125 ÷ .12½.
16. 25 ÷ .25.
17. .25 ÷ 25.
18. 1000 ÷ .001.
19. .001 ÷ 1000.
20. 18.65 ÷ 100.

95. To divide by 10, 100, 1000, etc., remove the point one place to the left for each cipher in the divisor.

Oral.

21. $38.64 \div 10$.
22. $.5 \div 10$.
23. $558 \div 100$.
24. $1684.32 \div 1000$.
25. $3.91 \div 1000$.
26. $1.155 \div 100$.
27. $398.42 \div 1000$.
28. $2.46 \div 200$.

NOTE. — To divide by 200, remove the point to the left, and divide by 2.

29. $386.54 \div 2000$.
30. $38.28 \div 400$.
31. $865.45 \div 5000$.
32. $2.5 \div 500$.

PARTS OF 100 OR 1000.

96. 1. What part of 100 is $12\frac{1}{2}$? 25? $33\frac{1}{3}$?
2. What part of 1000 is 125? 250? $333\frac{1}{3}$?
3. How much is $\frac{1}{5}$ of 100? Of 1000?
4. How much is $\frac{1}{4}$ of 100? Of 1000?
5. Find $\frac{1}{3}$ of 100. Of 1000.
6. How much is 25 times 24?

SOLUTION. — 100 times $24 = 2400$.
  25 times $24 = \frac{1}{4}$ as much as 100 times 24, $= 600$.

97. To multiply by 25, annex two ciphers, and take $\frac{1}{4}$ of the result.

7. Tell how to multiply by $33\frac{1}{3}$; by $12\frac{1}{2}$; by 250; by 125; by $333\frac{1}{3}$.

Oral.

8. 36×25.
9. $48 \times 12\frac{1}{2}$.
10. $24 \times 33\frac{1}{3}$.
11. 444×25.
12. $320 \times 33\frac{1}{3}$.
13. 125×80.
14. $333\frac{1}{3} \times 30$.
15. 168×250.
16. $12\frac{1}{2} \times 48$.

17. What cost 650 oysters at 50 cents a hundred?

SOLUTION. — $650 \div 100 = 6.50$ hundred.
  $\$.50 \times 6.50 = ?$

DECIMAL FRACTIONS.

18. What will be the cost of 3850 laths at 40 cents a hundred?

19. What is the freight on 685 pounds of baggage at $1.10 per 100 lb.

NOTE. — C. means 100 ; M., 1000.

20. What is the cost of 4862 ft. of pine lumber at $30 per M.?

21. Find the cost of 38,586 bricks at $8.25 a thousand.

22. What will 583 heads of cabbage cost at $3.50 a hundred?

23. At $3.50 a thousand, what will be the cost of 7800 shingles?

24. At $8.25 per C., what will be the cost of 2864 lb. of dried fish?

25. At $50 per M., what will be the cost of 3865 feet of cherry lumber?

26. What is the cost of laying 5890 bricks at $9.00 a thousand?

To find the cost of merchandise sold by the ton, divide the price by 2 and proceed as above.

27. Three loads of hay weigh 7894·lb. What will the hay bring at $12 a ton?

NOTE. — 1000 lb. will cost ½ of $12 = $6. $6 × 7.984 = ?

28. What cost 48986 lb. of railroad iron at $35 a ton?

29. Four loads of coal weigh respectively 3896 lb., 3524 lb., 4106 lb., and 3123 lb. What is the cost of the coal at $4.82 a ton.

ALIQUOT PARTS OF $1.00.

98. The **Aliquot Parts** of a number are the numbers which are exactly contained in it.

The aliquot parts of 100 are 5, 20, 12½, 16⅔, 33⅓, etc.

99. The aliquot parts of $1, commonly used, are as follows:

$6\frac{1}{4}$ cents = $\$\frac{1}{16}$.   25 cents = $\$\frac{1}{4}$.
$8\frac{1}{3}$ cents = $\$\frac{1}{12}$. $33\frac{1}{3}$ cents = $\$\frac{1}{3}$.
$12\frac{1}{2}$ cents = $\$\frac{1}{8}$.   50 cents = $\$\frac{1}{2}$.
$16\frac{2}{3}$ cents = $\$\frac{1}{6}$.

1. What is the cost of 69 books at $16\frac{2}{3}$¢ each?

SOLUTION. — 69 books will cost 69 times $16\frac{2}{3}$¢, or $69 \times \$\frac{1}{6} = \$\frac{69}{6} = \$11.50$. *Ans.*

100. Oral.

Multiply:

2. $33\frac{1}{3}$ cents by 36.   5. 25 cents by 40.
3. $12\frac{1}{2}$ cents by 24.   6. 75 cents by 4.
4. $6\frac{1}{4}$ cents by 32.

7. What is the cost of:
   48 lb. of bacon at $12\frac{1}{2}$¢ a pound?
   80 hand balls at 50¢ each?
   36 yd. of ribbon at $33\frac{1}{3}$¢ a yard?
   80 lb. of candy at 25¢ a pound?

101. Written.

8. Find the cost of the following:
   66 lb. of pork at $12\frac{1}{2}$¢.
   148 lb. of veal at $16\frac{2}{3}$¢.
   48 boxes of strawberries at 25¢.
   48 lb. of honey at 25¢.
   64 bars of soap at $6\frac{1}{4}$¢.
   60 doz. of eggs at $16\frac{2}{3}$¢.

Find the cost of:

9. 1580 lb. of sugar at $6\frac{1}{4}$¢ a pound.
10. 500 books at 25¢ each.
11. 16 yd. of dress-goods at $33\frac{1}{3}$¢ a yard.

DECIMAL FRACTIONS.

12. At 25¢ a pound, how many pounds of butter can be bought for $8.00 ?

SOLUTION. — As many pounds as 25¢ or $\frac{1}{4}$ is contained times in $8.00. $8 ÷ $\frac{1}{4}$ = 8 × $\frac{4}{1}$ = 32 lbs. *Ans.*

102. Oral.

Divide:

13. $5 by 33$\frac{1}{3}$¢.   16. $3 by 8$\frac{1}{3}$¢.
14. $6 by 6$\frac{1}{4}$¢.   17. $4 by 25¢.
15. $9 by 12$\frac{1}{2}$¢.   18. $4 by 66$\frac{2}{3}$¢.

19. At 25¢ each, how many hats can be bought for $6 ?

20. At $\frac{1}{4}$ a pound, how many pounds of cheese can be bought for $6 ?

21. At 33$\frac{1}{3}$¢ a yard, how many yards of linen can be bought for $10 ?

103. Written.

22. At 75¢ a bushel, how many bushels of barley can be bought for $125 ?

23. When butter is 25¢ a pound, how many pounds can I buy for $50 ?

24. How many dozen eggs at 16$\frac{2}{3}$ cents a dozen can be bought for $38 ?

25. At 12$\frac{1}{2}$ cents a quart, how many quarts of nuts can be bought for $10 ?

REVIEW OF DECIMALS.

104. 1. Tell how to locate the decimal point in any sum. In any remainder. In any product. In any quotient.

2. In the number 777, what is the local value of the 7 at the right ? The second 7 ? The left-hand 7 ?

3. Upon what does the value of any figure depend?

4. In the decimal .777, what is the value of the first 7 at the right? The second 7? The third 7?

5. What is the effect of removing an integral figure one place to the right? A decimal figure?

6. What is the effect of removing an integral figure one place to the left? A decimal figure?

Read:

7. .0001,   .00196,   4.3,
   .0006,   .02789,   71.86,
   .0014,   .52000,   329.400,
   .0282,   .050798,   1.001,
   .5897,   .725386,   200.3278,
   .00001,   .500001,   579000.00005,
   .00027,   .000829,   437.050609.

Copy and write decimally:

8. 1 tenth; 24 hundredths; 379 thousandths; 1000 ten-thousandths; 85 hundred-thousandths; 20079 millionths.

9. One thousand six and five hundred two millionths.

10. Three hundred fifteen thousand one, and eleven ten-thousandths; thirty-eight, and seven thousandths; 8 million 270 thousand 942, and 5 thousandths; seventeen tenths.

11. Four hundred 21, and 5 ten-thousandths; 1 thousand 27, and 27 hundredths; ninety-nine and ninety-nine ten-millionths.

Write without the denominator:

12. $\frac{1}{10}$; $\frac{2}{100}$; $\frac{3}{1000}$; $\frac{4}{10000}$; $\frac{5}{100000}$; $\frac{8}{1000000}$; $12\frac{17}{100}$; $42\frac{32}{100}$; $78\frac{589}{1000}$; $200\frac{2001}{10000}$.

13. Change to common fractions in lowest terms:

.028, .0015, .2175, .000048, .00075, .45, .8, .75, 8.9375, 91.16, 4001.645, 9.156575.

DECIMAL FRACTIONS.

Change to equivalent decimals:

14. $\frac{5}{8}$, $\frac{2}{8}$, $\frac{4}{25}$, $1\frac{5}{8}$, $\frac{3}{64}$, $\frac{3}{8}$, $20\frac{26}{47}$, $8\frac{1}{350}$, $4\frac{1}{16}$, $708\frac{131}{200}$.

Change to common fractions, then to simple decimals:

15. $.1\frac{1}{2}$, $.07\frac{3}{4}$, $.18\frac{1}{8}$, $.107\frac{7}{8}$, $.12\frac{1}{2}$, $.08\frac{1}{3}$, $.22\frac{2}{3}$, $.045\frac{2}{50}$, $.37\frac{3}{4}$, $.38\frac{2}{8}$, $.54\frac{5}{8}$, $.00005\frac{1}{4}$, $.78\frac{7}{8}$, $.38\frac{1}{8}$.

Reduce to a common denominator and add:

16. 50.06, 367.41, 200.200, .12304, 40.0056, 7.5620, .096071.

17. 1301.6, 904.02, .547, .0009, .00001, 218.94, 203.410, 1000, .01.

18. 100.101, 82.4, 401.009, .00038, 60702, 10.10, 574.68139.

19. 5.628, 850.002, 9.00256, 37.0005, 724.6811, 3759, 7000.0036, 2.25.

20. $11.78, $347, $5.06, $218, $20.07, $42.0244, $7.104, $37.625.

21. 4.76, .390, .0915, .00207, 841, 63.2, .00234, 1.43, .00536.

22. .00908, .0371, 24.5, 7.03, .0127, 354, .000781, .0436, 20.7354.

Subtraction.

23. $5.74 - 3.23 = ?$ 26. $367. - 1.52 = ?$

24. $.876 - .343 = ?$ 27. $200 - .02 = ?$

25. $67.5 - 41.5 = ?$

28. Which is greater, $\frac{3}{6}$ or 4 tenths?

29. How much more is $20 than $17.84?

30. From two million take two millionths.

31. I bought 4 farms; one contained 19.368 acres; one, 27.96 acres; one, 473.0008 acres; and the last one, 73.7561 acres. I sold 300.25 acres; how much land had I left?

32. From 1 inch take one ten-thousandth of an inch.

Multiply:

33. 7.945 by .3.
34. 350 by .42.
35. One tenth by one hundredth.
36. 25 units by 25 tenths.
37. 7.853 by 23.16.
38. $1.36 \times 20.04 = ?$
39. $27.27 \times 4.0004 = ?$

40. If wheat is worth $.38 a bushel, what will 117.75 bushels cost?

41. Apples sell for $1.28 a bushel; how much money will 24 barrels bring, each containing $2\frac{1}{2}$ bu.?

42. Find the cost of 3.325 lb. of butter at 18.75 cents a pound.

43. What will $6\frac{3}{4}$ yd. of broadcloth cost at $1.375 a yd.?

44. A boy paid $.125 a dozen for 1.75 dozen eggs; what did they cost him?

45. $3.64 \times .0002 \times 1.756 \times 4.004 = ?$

Divide:

46. 1738.89 by .00417.
47. 1237.6 by 26.
48. 36.11 by .021.
49. 2.38 by .17.
50. 36.82 by .0003.
51. 437.96 by 2.8.
52. 42.475681 by .29.
53. 40.20 by .000012.
54. $302.03 by 200.
55. 64.64006 by .002.
56. 12.9643 by 18.4.
57. 759.806 by 90.3.

58. $16\frac{3}{4} + 3.06 - \frac{5}{8} + .002 - 2.1 + .03 - \frac{3}{4} + .00\frac{1}{2} = ?$

59. $\frac{1}{8} + \frac{3}{4} - .65 + .5 + \frac{7}{8} - \frac{1}{6} + 3.14 = ?$

60. Find the product of .003 multiplied by .06, and divide it by 3.

61. A certain decimal divided by 1000 is 35.002. What is one fifteenth of the decimal?

ACCOUNTS AND BILLS. 25

62. The sum of two numbers is 306.52; one of them is 100. What is the other?

63. A man spent $450, which was .125 of his money. How much money had he?

64. Mr. A. bought a cow for $45, which was .375 of what he paid for a horse. How much did he pay for the horse?

65. John spent .75 of his money for a book and had $50 left. How much had he at first?

ACCOUNTS AND BILLS.

105. An **Account** is a record of indebtedness for articles bought or sold, cash paid or received, or services rendered.

106. A **Debtor** is a person who owes a debt.

107. A **Creditor** is a person to whom a debt is owed.

108. A **Bill** is a written statement of a debtor's account, made by the creditor.

109. A **Receipt** is a creditor's written acknowledgment that he has received payment of part or all of a debt.

110. A bill is receipted when its payment is acknowledged in writing, by the creditor, or by some authorized person.

NOTE.— The sign @ is for at. Dr. is for debtor, and Cr. for creditor.

BILL FORMS.

SYRACUSE, N.Y., *July 1, 1896.*

JAMES P. BARNES, *Bought of* DEY BROS. & Co.

50 yd. Brussels Carpet	@	$1	15	$
24 " Oil Cloth	"		35	
4 doz. pair Merino Hose	"	3	50	
2 Willow Chairs	"	4	50	
				$

SENIOR ARITHMETIC.

RECEIPTED BILL WITH CREDITS.

ROCHESTER, N. Y., *Jan. 2, 1896.*

MRS. JOHN F. WHITE,

1895   *To* BURKE & WHITE, *Dr.*

Nov.	6	4 lb. Coffee	@	$	27	$	
"	6	28 lb. Sugar	"		5½		
"	18	5 gal. Molasses	"		60		
Dec.	11	18 lb. Rice	"		7¼		
"	15	2 bbl. Potatoes	"	1	80		
"	19	28 lb. Butter	"		21		
		CR.					
Nov.	18	Cash		3	50		
Dec.	28	"		4	75		
		Balance due,					

Received payment, *Jan. 15, 1896,*

     BURKE & WHITE,

       By JOHN R. PIERCE.

FORM OF A RECEIPTED BILL.

NEW YORK, *June 30, 1896.*

JEROME A. PHELPS,

   In account with D. O. POTTER & Co.

May	14	12 bbl. Flour	@	$6	50	$	
"	14	6 tubs Butter, 684 lbs.	"		24		
June	10	5 bbl. Beef	"	25	28		
"	25	450 lb. Ham	"		9¼		

     Received payment,

       D. O. POTTER & Co.

INDICATED OPERATIONS.

111. The **Parenthesis**, ( ), indicates that all the numbers contained therein are to be taken together.

112. Brackets, [ ], **Braces,** { }, and the **Vinculum,** ‾‾‾, have the same use as the parenthesis.

113. When the parenthesis is not used, operations indicated by × or ÷ must be performed first. Thus,

1. $12 \div 4 \times 2 + 36 \div 4 - 2 \times 4 = ?$

SOLUTION. —

$12 \div 4 \times 2 = 6.$
$36 \div 4 = 9.$ $\qquad 6 + 9 - 8 = 7.$ *Ans.*
$2 \times 4 = 8.$

2. $4 + 3 \times 2 = ?$ $\qquad$ 5. $4 \times (3 + 2) = ?$
3. $(4 + 3) \times 2 = ?$ $\qquad$ 6. $8 + 4 \div 2 = ?$
4. $4 \times 3 + 2 = ?$ $\qquad$ 7. $(8 + 4) \div 2 = ?$

NOTE. — When one parenthesis, brace, or vinculum includes another, first remove the inner one.

114. Find the value of:

1. $15 + 3 \times 6 + 10 \div 5.$
2. $(6 + 4) \times (3 + 2) - (8 \times 5).$
3. $18 \div 3 \times 2 + 8 \times 2 \div 4 - 6.$
4. $2 + 12 \div 4 - (10 + 6 \div 4) \div 3.$
5. $(11 + 4) \div 3 + 6 \times 4.$
6. $3 + 4 \times 6 \div (15 + 9 \div 3).$
7. $164 + 16 - 250 \div 10 + 16 \times 3.$
8. $17 + 3 \times 4 \times 6 + 3 \div 3 + 3.$
9. $[39 + 8 \div 2 + 7] \times 6.$
10. $[6 + 15 \times 3 - (6 + 16 \div 8 + 4)] \div 8 + 5.$

MISCELLANEOUS.

115. 1. I have four pieces of broadcloth. The first contains 13.7642 yd.; the second, 22.008 yd.; the third, 15.027 yd.; and the fourth, 19.255 yd. How many yards in all?

2. From a piece of ribbon containing $103\frac{1}{2}$ yd., $73\frac{3}{4}$ yd. were sold. How many yards were left?

3. How many yards of muslin at $\$.12\frac{1}{2}$ a yard will it take for 4 pair of curtains, if each curtain contains 3.375 yd.?

4. I have 14.735 yd. of lace, and desire to cut it into seven equal strips; how much will there be in each strip?

5. What will be the cost of a hat at $7.50, a pair of gloves at $1.13, a veil at $1.25, and a parasol at $3.375?

6. Arrange the following articles in the form of a bill: 7 qt. of molasses at $.15 a qt., 3 pk. of apples at $1.28 a bushel, 30 lb. of sugar at $.08$\frac{1}{2}$ a pound, and 12 bu. of potatoes at $.29 a bushel.

7. A grocer bought three bunches of bananas at $1.54 a bunch. The first bunch contained 73 bananas, the second 54, and the third 97. He sold them all at 30¢ a dozen; did he gain or lose, and how much?

8. The first year in business a grocer made $2374.68, the second $1529.47, and in the third year he lost $300. His expense each year averaged $928.45; how much money had he gained at the end of three years?

9. What will 9 barrels of flour cost, if 28 barrels cost $173.60?

10. I bought 437 heads of lettuce at $5 a hundred, and sold them at $.08 a head; what was my gain?

MISCELLANEOUS.

Find the cost of:

11. 6824 lb. of coal at $4.68 a ton.
12. 2384 lb. of coal at $5.67 a ton.
13. 8972 ft. of lumber at $35.40 a thousand.
14. 6854 lb. of hay at $16.50 a ton.
15. 4836 bricks at $9.45 per M.
16. 895 ft. of lumber at $19.75 per M.
17. What part of 4.50 is $3.33\frac{1}{3}$?
18. What part of 3.625 is 1.5?
19. What part of 6.2 is 3.25?
20. 1.1 is what part of 7.4?

21. A father left his son $24000, which was .375 of his estate. What was the value of the estate?

22. Divide 26 by $2\frac{1}{3}$, and multiply the result by 17.345.

23. Divide $\frac{3}{4}$ of .375 by $\frac{5}{8}$ of $\frac{2}{3}$ of .298.

24. The product of three numbers is 167.7. Two of the numbers are 3.25 and 5.16. What is the other?

25. What number divided by 2.86 equals .34?

26. What number diminished by 38.64 leaves .356?

27. A man bought 8.5 yd. of cloth at 3.33\frac{1}{3}$ a yard, 12.4 yd. at $2.75, $18\frac{1}{3}$ yd. at $4.375, and $24\frac{5}{8}$ yd. at $2.875. How many bushels of corn at $43\frac{3}{4}$ cents a bushel will pay for the cloth?

28. .5 of a number exceeds .45 of it by 20. What is the number?

SOLUTION: $.5 - .45 = .05$. Now the question is, 20 is .05 of what? $20 \div .05 = 400$.

29. At 85¢ a yard, how many yards of cloth can be purchased for $29.75.

30. Divide $785 among A, B, and C, so that C will have $185 more than each of the others.

31. $\dfrac{1}{.05} - \dfrac{.0045}{.4 \times .005 + .002 \times .125} = ?$

32. What part of .876 is .31536?

33. If .375 of a ton of coal cost $1.25, what will 7.125 tons cost?

34. What is .3 of a number when .8 of it is 80?

35. How many thousandths in 3 units?

36. How many thousandths in .1?

37. Express $\frac{1}{4}$ of one hundredth as a decimal.

38. The salary of the President of the U. S. is $50,000 a year. How much does he receive per day?

116. Review of Factors, Multiples, Divisors, and Cancellation.

1. Define factor, composite number, prime number, and prime factor.

2. Find the prime factors of 5075; of 9576; of 3150; of 6006.

3. Find the sum of the prime factors of 34650.

4. Find the prime factors of 2310; of 17199; of 6840.

5. 81158 is the product of what prime factors?

6. Find the largest prime factor of 12600.

7. What is a common divisor of two or more numbers?

8. What is the greatest common divisor of two or more numbers?

9. When are numbers prime to each other.

Find the greatest common divisor of:

10. 672 and 960.
11. 616 and 1012.
12. 272 and 428.
13. 1650 and 1920.
14. 696, 1218, and 1160.
15. 450, 720, and 810.

MISCELLANEOUS.

16. What is the greatest prime factor common to 4242 and 2626.

17. A grocer had 84 bananas and 126 lemons, which he wished to put into bags, each bag containing the largest number possible, and each containing the same number. How many could be put into each bag?

18. A man has three fields containing respectively 14, 18, and 22 acres. He wishes to cut them into the largest possible lots of equal size. How much land will each lot contain? How many lots will each field contain?

19. What is a multiple of a number? A common multiple? The least common multiple?

Find the least common multiple of:

20. 96, 196, 42, and 54.   **23.** 252, 462, and 1092.

21. 45, 36, 70, and 90.   **24.** 120, 280, and 308.

22. 36, 40, 42, and 48.   **25.** 36, 110, 98, and 66.

26. Find the least common multiple of the even numbers to and including 20.

27. What is the least sum with which I can buy an exact number of chairs at $6, $8, or $5 each?

28. What is the smallest sum of money that may be expended by using an exact number of nickels, dimes, quarters, or 3-cent pieces?

How many pieces of each kind will the sum contain?

29. John can run around a block in 6 minutes, James in 8 minutes, and Henry in 9 minutes. If they start together, how long before they will all be together again at the starting-point?

30. What is the shortest piece of rope that can be cut into pieces 32, 36, and 44 feet long?

31. What is cancellation?

32. Of what use is cancellation?

Find results of the following by cancellation:

33. $\dfrac{18 \times 36 \times 48}{24 \times 6 \times 12}$

34. $\dfrac{10 \times 6 \times 4}{4 \times 6}$

35. $\dfrac{18 \times 9 \times 10 \times 5}{6 \times 8 \times 2}$

36. $\dfrac{28 \times 56 \times 30}{14 \times 3 \times 5}$

37. $\dfrac{28 \times 32 \times 7}{14 \times 35 \times 2}$

38. $\dfrac{34 \times 9 \times 5}{25 \times 17 \times 3}$

39. $240 \times 48 \times 70 \times 18 \div 42 \times 15 \times 54 \times 7 = ?$

40. Divide the product of 25, 14, and 11 by the product of 15, 7, and 22.

41. How many bushels of wheat at $1.10 a bushel must be given for 6 pieces of cloth each containing 33 yards at 50 cents a yard?

42. How many cords of wood at $3 a cord will pay for 30 lb. of sugar at 5 cents a pound?

43. If 8 men can do a piece of work in 6 days, in how many days can 12 men do it?

44. How many pounds of sugar can be bought for $7 if 21 lb. cost $1.05?

45. How many pounds of maple sugar at 12 cents a pound must a farmer exchange for 15 pounds of coffee at 24¢ a pound?

46. A milkman exchanges 8 cans of milk, 30 quarts in a can, at 4 cents a quart, for 3 pieces of sheeting, 40 yards in a piece. What is the price of the sheeting per yard?

MISCELLANEOUS REVIEW OF COMMON FRACTIONS.

117. Oral.

1. Add $\frac{1}{4}$ and $\frac{1}{3}$; $\frac{1}{2}$ and $\frac{1}{8}$; $\frac{2}{3}$ and $\frac{3}{4}$; $\frac{6}{7}$ and $\frac{5}{8}$; $\frac{3}{4}$ and $\frac{7}{8}$; $\frac{7}{8}$ and $1\frac{1}{2}$.

2. $\frac{1}{2} - \frac{1}{4} = ?$ $\frac{1}{3} - \frac{1}{5} = ?$ $\frac{3}{4} - \frac{3}{8} = ?$ $1\frac{3}{4} - \frac{7}{8} = ?$

MISCELLANEOUS REVIEW OF COMMON FRACTIONS. 33

3. Reduce to improper fractions $3\frac{1}{4}$, $7\frac{3}{8}$, $8\frac{2}{3}$, $5\frac{1}{3}$, $7\frac{6}{7}$, $8\frac{5}{8}$, $16\frac{3}{9}$, $15\frac{1}{4}$.

4. Reduce to integers or mixed numbers $\frac{14}{4}$, $\frac{17}{9}$, $\frac{14}{6}$, $\frac{11}{5}$, $\frac{40}{3}$, $\frac{16}{3}$, $\frac{85}{9}$, $\frac{132}{11}$, $\frac{150}{12}$.

5. Multiply 16 by $\frac{3}{8}$; 45 by $\frac{3}{9}$; 18 by $\frac{5}{6}$; 45 by $\frac{3}{15}$; $\frac{2}{3}$ by 9; $\frac{7}{8}$ by 32; $1\frac{5}{8}$ by 16; $\frac{7}{9}$ by 27.

6. Find product of: $\frac{3}{4} \times \frac{6}{7}$; $\frac{8}{11} \times \frac{5}{12}$; $3\frac{1}{2} \times 1\frac{1}{4}$; $\frac{3}{8} \times 1\frac{5}{7}$; $\frac{6}{7} \times 2\frac{1}{4}$; $4\frac{1}{2} \times 6\frac{2}{3}$.

7. Find $\frac{2}{3}$ of 24; $\frac{3}{8}$ of 12; $\frac{5}{6}$ of 30; $\frac{7}{9}$ of 27; $\frac{4}{5}$ of 45; $\frac{7}{8}$ of 40.

8. Find $\frac{1}{2}$ of $\frac{1}{3}$; $\frac{1}{6}$ of $\frac{3}{8}$; $\frac{2}{7}$ of $\frac{3}{7}$; $\frac{7}{10}$ of $\frac{6}{8}$; $\frac{5}{8}$ of $1\frac{6}{7}$; $\frac{1}{3}$ of $2\frac{1}{2}$.

9. Divide $\frac{6}{8}$ by 3; $\frac{4}{5}$ by 4; $\frac{8}{12}$ by 12; $\frac{8}{11}$ by 11; $4\frac{1}{2}$ by 3; $\frac{7}{8}$ by 6; $4\frac{2}{3}$ by 6; 10 by $\frac{5}{7}$; 8 by $\frac{7}{8}$; $1\frac{5}{6}$ by 8; $\frac{3}{11}$ by 22; $\frac{4}{7}$ by 9.

10. Divide 4 by $\frac{1}{3}$; 8 by $\frac{1}{5}$; 9 by $\frac{3}{7}$; 16 by $\frac{2}{5}$; 24 by $\frac{5}{6}$; 13 by $\frac{2}{3}$; 11 by $\frac{4}{5}$; 12 by $1\frac{2}{3}$.

11. Divide $\frac{1}{2}$ by $\frac{3}{4}$; $\frac{5}{6}$ by $\frac{2}{3}$; $\frac{1}{2}$ by $\frac{1}{4}$; $\frac{3}{8}$ by $\frac{4}{5}$; $\frac{7}{10}$ by $\frac{5}{8}$; $\frac{6}{7}$ by $\frac{4}{5}$; $5\frac{1}{2}$ by $2\frac{1}{3}$.

12. Divide 1 by:

$\frac{1}{3}$, $\frac{1}{5}$, $\frac{1}{6}$, $\frac{1}{15}$, $\frac{1}{7}$, $\frac{1}{25}$, $\frac{1}{10}$, $\frac{3}{4}$, $\frac{2}{3}$.

NOTE. — 1 divided by a fraction equals that fraction inverted.

13. $\frac{3}{4}$ of 12 = ?   9 = what part of 12 ?   9 is $\frac{3}{4}$ of what ?

14. $3 \times ? = \frac{2}{3}$; $\frac{5}{8} \div ? = \frac{1}{2}$; $\frac{3}{4} \times ? = \frac{1}{2}$; $\frac{5}{6} \div ? = \frac{1}{8}$; $\frac{3}{5} \div 2 = \frac{1}{5}$; $\frac{5}{8} \div ? = \frac{8}{5}$.

15. $\frac{1}{2} - ? = \frac{1}{3}$; $\frac{1}{5} + ? = \frac{1}{3}$; $\frac{2}{3} - ? = \frac{2}{5}$; $\frac{5}{8} + ? = 1\frac{1}{2}$; $\frac{4}{5} - ? = \frac{1}{2}$.

16. What part of
   6 is 4 ? $\frac{1}{2}$ is $\frac{1}{4}$? $\frac{3}{4}$ is $\frac{2}{8}$? $5\frac{1}{2}$ is $2\frac{1}{4}$?
   11 is 5 ? $\frac{1}{3}$ is $\frac{1}{6}$? $\frac{5}{6}$ is $\frac{1}{5}$? $\frac{2}{3}$ is $\frac{1}{3}$?

17. 9 is $\frac{2}{3}$ of what ?   5 is $\frac{5}{8}$ of what ?   6 is $\frac{3}{8}$ of what ?

18. Change $\frac{5}{8}$ to 24ths; $\frac{2}{3}$ to 15ths; $\frac{7}{8}$ to 32nds; $\frac{3}{4}$ to 20ths.

19. A man owning $\frac{3}{4}$ of a farm, sold $\frac{1}{3}$ of his share. What part did he sell? How much remains?

20. At $12\frac{1}{2}$¢ a dozen, how many dozen of eggs can I buy for $3?

21. At $6\frac{1}{4}$¢ a box, how much will 8 boxes of berries cost?

22. John has 56 cents, and James $\frac{1}{8}$ as much. How much have both?

23. A can do a piece of work in 4 days; B can do the same piece of work in 2 days. What part of the work can each do in a day?

24. A can mow a field in 3 days, and B in 4 days. What part of the field can they mow in a day if both work together?

A can mow $\frac{1}{3}$ of it in 1 day, and B can mow $\frac{1}{4}$ of it in 1 day. Both working together can mow the sum of $\frac{1}{3}$ and $\frac{1}{4} = \frac{7}{12}$ of it in 1 day.

25. C can do a piece of work in 2 days and D can do it in 4 days. In what time can they both do it, working together?

C does $\frac{1}{2}$ of it in 1 day, and D $\frac{1}{4}$ of it in 1 day. Therefore both can do $\frac{1}{2} + \frac{1}{4} = \frac{3}{4}$ of it in 1 day. Since both can do $\frac{3}{4}$ of it in 1 day, it will take as many days to do $\frac{4}{4}$ or the whole of it, as $\frac{3}{4}$ is contained times in $\frac{4}{4}$, or $1\frac{1}{3}$ days. *Ans.*

NOTE.— $\frac{4}{4}$ divided by $\frac{3}{4}$ gives the same result as 4 divided by 3.

26. $\frac{1}{3}$ of my money is gold, and $\frac{1}{2}$ as much is silver. What part of my money is silver?

27. If a boy can earn 2\frac{1}{2}$ in 1 week, how much can 3 boys earn in 4 weeks?

28. James sold a book for 28 cents, which was $\frac{2}{3}$ of what it cost him. What did it cost him?

MISCELLANEOUS REVIEW OF COMMON FRACTIONS.

29. The difference between $\frac{1}{2}$ of a number and $\frac{1}{4}$ of it is 6. What is the number?

SOLUTION. — The difference between $\frac{1}{2}$ and $\frac{1}{4}$ is $\frac{1}{4}$. Now the question is 6 is $\frac{1}{4}$ of what?

30. A boy 12 years of age is $\frac{1}{4}$ as old as his father. How old is his father?

Written.

31. A farmer having 1200 bushels of potatoes, sold $\frac{1}{8}$ of them at one time, $\frac{1}{4}$ at another, and 350 bushels at another. How many bushels had he left?

32. A mechanic whose wages are $5 per day uses $\frac{3}{10}$ of his weekly earnings for board, and $\frac{2}{5}$ for clothing and other expenses. How many dollars does he save weekly?

33. Which is greater and how much, $1\frac{3}{4}$ or $1\frac{4}{5}$?

34. If it takes 27 days to do a piece of work, how long will it take to do $\frac{2}{3}$ of it?

35. If a horse is worth $100, and a cow is worth $\frac{2}{5}$ as much as the horse, what is the cow worth?

36. John has in the bank $45 and draws out $\frac{2}{5}$ of it. How much remains in the bank?

37. What will 16 pair of shoes cost at 3\frac{7}{8}$ a pair?

38. If a farmer has 23 sheep and sells them at 3\frac{9}{10}$ apiece, how much does he receive for the sheep?

39. What is the cost of $\frac{9}{10}$ of a pound of cheese at 10¢ a pound?

40. What is the cost of $\frac{8}{9}$ of a yard of cloth at 1\frac{1}{4}$ a yard?

41. $\frac{8}{9} \times \frac{4}{5} \times 6\frac{1}{2} \times \frac{4}{15} \times 3 \times 1\frac{8}{9} = ?$

42. What is the value of $3\frac{1}{4}$ of $8\frac{7}{8}$ of $\frac{2}{3}$ of $1\frac{8}{35}$?

43. A man sold $3\frac{2}{3}$ tons of hay at one time, $7\frac{2}{5}$ at another, and enough the third time to make 20 tons. How many tons did he sell the third time?

44. $\frac{1}{12}$ plus $\frac{1}{6}$ plus $\frac{1}{4}$ plus $\frac{1}{3}$ and how many more will make 3?

45. A man having a farm of 96 acres, sold $\frac{1}{2}$ of an acre to one man, $\frac{1}{3}$ of an acre to another, $\frac{2}{3}$ of an acre to another, and $\frac{1}{15}$ of an acre to another. How many acres had he left?

46. If two men were 90 miles apart and each should travel $23\frac{1}{2}$ miles toward the other, how many miles would they then be apart?

47. If George has $\frac{1}{2}$ a dollar and $\frac{1}{5}$ of a dollar, and Henry has $\frac{1}{4}$ of a dollar and $\frac{3}{10}$ of a dollar, which has the greater amount and how much?

48. A man bought 3 loads of wood containing respectively $1\frac{1}{4}$ cords, $1\frac{3}{8}$ cords, and $1\frac{5}{8}$ cords. How many cords of wood did he buy?

49. I paid $\$10\frac{1}{2}$ for hay, $\$15\frac{2}{3}$ for coal, and $\$6\frac{1}{4}$ for wood. What did I pay for all?

50. Mr. Jones paid $\$525\frac{1}{4}$ for a span of horses, and sold them for $\$625\frac{5}{8}$. How much did he gain?

51. L. W. and J. E. Connell paid $\$4500\frac{7}{8}$ for a store and its contents. They sold it for $\$5025\frac{3}{4}$. How much did they gain by the operation?

52. A, B, C, D, and E own respectively $\frac{1}{2}$, $\frac{2}{3}$, $\frac{5}{8}$, $\frac{9}{10}$, and $1\frac{1}{2}$ acres of land. How much do they all own?

53. A gentleman having $\$1700$, paid $\$825\frac{1}{4}$ for horses, $\$230\frac{2}{3}$ for cows, $\$150\frac{7}{8}$ for oxen, and $\$407\frac{5}{8}$ for sheep. How much money had he left?

54. Mr. Blanchard paid $\$8\frac{9}{10}$ for shovelling his walk, $\$5\frac{2}{3}$ for trimming his grape-vines, and $\$6\frac{3}{4}$ for sifting his ashes. He gave the man a 20-dollar bill and a dollar bill. How much money should Mr. B. receive in return

MISCELLANEOUS REVIEW OF COMMON FRACTIONS. 37

55. If I add 2 to each term of the fraction $\frac{1}{3}$, will its value be increased or diminished, and how much?

56. Mr. Homer has $10\frac{1}{2}$ acres of wheat, $6\frac{3}{4}$ acres of corn, $20\frac{3}{8}$ acres of barley, and $16\frac{3}{8}$ acres of rye. How many acres of grain has he?

57. What is the quotient of 389 divided by 1556, expressed in its simplest form?

58. $\dfrac{(\frac{3}{4} + \frac{5}{6}) \div \frac{4}{5}}{\frac{2}{5} \text{ of } 1\frac{2}{3} \text{ of } \frac{9}{11} \text{ of } \frac{2}{3}} = ?$

59. 816 is $\frac{4}{5}$ of what number?

60. From $\frac{3}{5}$ of $1\frac{9}{8}$ take $\frac{2}{7}$.

61. The product of two factors is $10\frac{1}{2}$; one factor is $3\frac{2}{5}$. What is the other?

62. $\frac{3}{8} + 6\frac{2}{7} + 9\frac{1}{4} + \frac{6}{12} + \frac{7}{16} = ?$

63. $(\frac{4}{15} \text{ of } 2\frac{1}{2} \text{ of } \frac{3}{17}) \times (\frac{2}{9} \text{ of } 3\frac{1}{4} \text{ of } 8 \text{ of } \frac{1}{7}) = ?$

64. The sum of two numbers is $19\frac{4}{13}$. One of the numbers is $12\frac{3}{7}$. What is the other?

65. Reduce $\dfrac{4\frac{3}{8}}{28}$ to its lowest terms.

66. $(\frac{7}{8} - \frac{3}{5}) \times (\frac{4}{5} + \frac{2}{3}) = ?$

67. Change to simple fractions:

$\dfrac{1\frac{5}{6}}{9}, \dfrac{3\frac{1}{4}}{5\frac{1}{2}}, \dfrac{\frac{1}{2} \text{ of } \frac{5}{7}}{16}, \dfrac{7\frac{1}{2}}{\frac{3}{4} \text{ of } 2\frac{1}{2}}, \dfrac{\frac{3}{5} \text{ of } \frac{2}{3}}{\frac{1}{8} \text{ of } \frac{7}{8}}, \dfrac{3\frac{5}{6}}{1\frac{2}{5}}, \dfrac{\frac{3}{5} \text{ of } \frac{8}{9}}{\frac{1}{4} \text{ of } \frac{5}{8}}.$

68. A can do a piece of work in 4 days, B can do the same work in 5 days, and C in 6 days. In what time can all do it together?

69. A tank has 3 supply pipes. It can be filled in 6 hours by the first pipe, in 7 hours by the second, and in 8 hours by the third. In how many hours can the tank be filled by the three pipes together?

70. A and B can do a piece of work in 3 days. A can do it alone in $5\frac{1}{2}$ days. In what time can B do it alone?

SOLUTION. — Both can do $\frac{1}{3}$ of it in 1 day. A, alone, can do $\frac{2}{11}$ of it in 1 day. $\frac{1}{3} - \frac{2}{11} = \frac{5}{33}$, the part A can do in 1 day. Since he can do $\frac{5}{33}$ of it in 1 day, he can do $\frac{33}{33}$, or the whole of it, in as many days as $\frac{5}{33}$ is contained times in $\frac{33}{33}$, or $33 \div 5 = 6\frac{3}{5}$ days.

71. $\frac{1}{5}$ of my property is invested in land, $\frac{2}{7}$ of the remainder in business, and $\frac{2}{3}$ of the remainder, which is $2400, is in the bank. How much property have I?

72. What is the value of $\left(1\frac{2}{3} + 6 - \frac{2}{3} \text{ of } \frac{5}{6} + \frac{\frac{7}{8}}{12}\right) \div 3\frac{1}{5}$?

73. A farmer sold 11 doz. eggs at $14\frac{1}{2}$¢ a dozen, and took his pay in sugar at $5\frac{1}{2}$¢ a pound. How much did he receive?

74. Find the value of $\dfrac{3\frac{3}{4}}{\frac{2}{3} \text{ of } \frac{5}{6}} + \frac{1}{2} \text{ of } \frac{1}{3} \div \frac{2}{3}$.

75. A boy having spent $\frac{1}{2}$ of $\frac{2}{3}$ of his money for a knife, had $2.25 left. How much did he pay for the knife?

76. A father left $39000 to his two children, dividing it so that the daughter received $\frac{4}{5}$ as much as the son. What was the share of each?

77. A person owning $\frac{2}{3}$ of a steamboat, sold $\frac{4}{5}$ of his share for $17360. What was the value of the boat?

78. After spending $\frac{1}{2}$ of my money and $\frac{1}{4}$ of the remainder, I had $300 left. How much had I at first?

79. If $\frac{1}{2}$ of $\frac{5}{6}$ of a bushel of apples cost $\frac{3}{4}$ of $\frac{9}{10}$ of a dollar, what will $\frac{2}{3}$ of $\frac{4}{5}$ of a bushel cost?

80. How many pounds of honey at $\frac{1}{2}$ of $\frac{2}{7}$ of a dollar a pound can be bought for $\frac{5}{8}$ of $2\frac{3}{4}$ dollars?

81. Simplify $\dfrac{\frac{2}{3}}{\frac{1}{2} \text{ of } \frac{5}{6}} \div \dfrac{\frac{2}{3} \text{ of } \frac{4}{5}}{5\frac{1}{2}}$.

82. $\dfrac{\frac{2}{3} \text{ of } \frac{5}{8}}{12} \div \dfrac{\frac{5}{8} \text{ of } \frac{6}{16}}{2 \times 3\frac{1}{3}} = $?

QUESTIONS.

83. Divide the product of 5 times $\frac{3\frac{3}{8}}{4\frac{2}{7}}$ plus $\frac{\frac{1}{2} \text{ of } \frac{3}{8}}{15}$ by $\frac{5\frac{3}{4}}{4\frac{1}{2}}$.

84. Divide $\frac{2}{7}$ of $\frac{4\frac{2}{3}}{7\frac{1}{8}}$ by $\frac{7}{8}$ of $\frac{5\frac{1}{4}}{4\frac{1}{2}}$.

Find the cost of the following:

85. 315½ lb. of tea at $.37½ a pound.
    34¾ lb. of coffee at $.18¾ a pound.
    3105¾ lb. of pork at $.12½ a pound.
    3069½ bu. of wheat at $1.12½ a bushel.
    36¾ doz. of eggs at $.12½ a dozen.
    26¾ yd. of sheeting at $.07¾ a yard.

86. A owns ⅗ of a farm and B owns the remainder; ¾ of the difference of their shares is worth $2400. What is the value of the farm?

87. Divide $3½ among some poor children, giving each ⅛ of a dollar. What will be the number of children?

88. Two men hire a pasture for $25. A puts in 8 horses and B 12 horses. How much should each pay?

89. Add 8 to both terms of the fraction ⁶⁄₉, and find how much you have increased or diminished it.

90. Subtract 4 from each term of the fraction ⁶⁄₉ and find how much it has been increased or diminished.

QUESTIONS.

118. 1. What is a decimal? How are decimals written? Why are they called decimals?

2. How many decimal places are needed to write ten-thousandths? Millionths? Hundredths?

3. What is the first place at the right of the decimal point? What is the first period called? The second place? The second period?

4. What is a mixed decimal?

5. What must the denominator of a decimal be?

6. What is the effect of removing the decimal point one place to the right? To the left? Two places to the right? Three places to the left?

7. What is the effect of annexing a cipher to an integer? To a decimal? Of prefixing a cipher to an integer? To a decimal?

8. How do we reduce decimals to common fractions? Common fractions to decimals?

9. Give rules for adding, subtracting, multiplying, and dividing decimals.

10. How do we locate the decimal point in the sum? In the remainder? In the product? In the quotient?

11. What are coins?

12. What are the gold, silver, bronze, and nickel coins used in the U. S.?

13. What are the aliquot parts of a number? What are the aliquot parts of $1? Of 100? Of 1,000?

14. What is a bill? An account? A creditor? A debtor? Tell how to receipt a bill.

119. 1. Define unit, number, the unit of a number, abstract number, concrete number, like numbers.

2. Define notation, numeration, Arabic notation.

3. What is the value of the unit figure of a number? The tens? The hundreds?

4. What is the largest number which can be expressed by four figures?

5. What is the simple value of a figure? The local value?

6. What name is given to the first period at the right of the decimal point? The second? The third?

7. What is addition? What kind of numbers can be added?

8. Define subtraction, minuend, subtrahend, remainder. What is a proof of subtraction? What is the sign of subtraction, and where placed?

9. What is a parenthesis? A vinculum? For what are they used?

10. What is multiplication? The multiplier? The multiplicand? The product?

11. The multiplier and the multiplicand are what of the product?

12. What is the sign of multiplication and how is it used? Define division, divisor, dividend, quotient, remainder.

13. What is the sign of division, and how is it used?

14. Express the division of 12 by 8 in as many ways as you can.

15. To what terms in multiplication do the divisor, quotient, and dividend correspond?

16. How do you find the dividend when the divisor, quotient, and remainder are given?

17. When is the quotient an abstract number?

18. When the quotient and dividend are like numbers, what kind of a number is the divisor?

19. How can we divide when the divisor is 10? 100? 1000? When the divisor is 20? 50? 300?

20. Multiplying both dividend and divisor by the same number affects the quotient how?

21. Dividing both divisor and dividend by the same number affects the quotient how?

22. Multiplying the dividend affects the quotient how? The divisor? Dividing the dividend? The divisor?

23. Define exact divisor, factor, prime factor, factoring.

24. How can you find the prime factors of a number?

25. Define divisor. Common divisor. The greatest common divisor. Give rule to find greatest common divisor.

26. Define multiple, common multiple, least common multiple. Give rule for finding the least common multiple.

120. Define fraction, fractional unit, unit of a fraction, denominator, numerator, terms of a fraction, common fraction, integer, proper fraction, improper fraction, mixed number, simple fraction, compound fraction, complex fraction.

What is the value of a fraction?

State the principles of fractions.

What is it to reduce a fraction?

How are fractions reduced to lowest terms? To highest terms?

How can an improper fraction be reduced to a whole or a mixed number? A whole or a mixed number to an improper fraction?

What are like fractions? Unlike fractions?

How can fractions be reduced to others having a common denominator? A least common denominator?

How can two or more fractions be added?

How can the sum of fractions be found? Mixed numbers?

How can the difference of fractions be found? Mixed numbers?

How can a fraction be multiplied by a fraction? A fraction by an integer?

COMPOUND NUMBERS. 43

How can an integer be multiplied by a fraction? By a mixed number?
How can a fraction be divided by a fraction?
How do you reduce a complex fraction to a simple fraction?

COMPOUND NUMBERS.

121. A number composed of only one kind of unit is a **Simple Number**; as, 5 pk., 4 apples, 6.

122. A **Denomination** is a name given to a unit of measure or of weight.

123. A number composed of different kinds of units is a **Compound Number**; as, 3 bu. 2 pk. 1 qt.

A number having one or more denominations is also called a **Denominate Number**.

124. Reduction is the process of changing a number from one denomination to another without changing its value.

125. Changing to a lower denomination is called **Reduction Descending**; as, 2 bu. 3 pk. = 88 qt.

126. Changing to a higher denomination is called **Reduction Ascending**; 88 qt. = 2 bu. 3 pk.

127. Linear Measure is used in measuring lines or distances.

TABLE.

12 inches (in.) = 1 foot, ft.
3 feet = 1 yard, yd.
5½ yards, or 16½ feet = 1 rod, rd.
40 rods = 1 furlong, fur
320 rods, or 5280 feet = 1 mile, mi.
1 mi. = 320 rd. = 1760 yd. = 5280 ft. = 63360 in.

128. Surveyors' Measure is used in measuring land.

TABLE.

7.92 inches = 1 link, li.
100 links = 1 chain, ch.
80 chains = 1 mile, mi.

NOTE. — A surveyors' chain is 4 rods long, and contains 100 links. A chain, or steel measuring tape, 100 feet long, is sometimes used by engineers.

129. Square Measure is used in measuring surfaces.

TABLE.

144 square inches = 1 square foot, sq. ft.
9 square feet = 1 square yard, sq. yd.
$30\frac{1}{4}$ square yards ⎫
$272\frac{1}{4}$ square feet ⎭ = 1 square rod, sq. rd.
160 square rods = 1 acre, A.
640 acres = 1 square mile, sq. mi.

1 sq. mi. = 640 A. = 102400 sq. rd. = 3097600 sq. yd.

130. A square mile of land is called a **Section**.

A square rod is sometimes called a perch (P.). A rood (R.) is 40 sq. rods.

NOTE. — 1 acre = 43560 sq. ft. There are 10 square chains in an acre.

Roofing, paving, etc., are often estimated by the Square, which is 100 square feet.

131. Cubic Measure is used in measuring volumes or solids.

TABLE.

1728 cubic inches = 1 cubic foot, cu. ft.
27 cubic feet = 1 cubic yard, cu. yd.
16 cubic feet = 1 cord foot, cd. ft.
8 cord feet, or 128 cu. ft. = 1 cord, C.
1 cu. yd. = 27 cu. ft. = 46656 cu. in.

COMPOUND NUMBERS. 45

132. Liquid Measure is used in measuring liquids.

TABLE.

4 gills (gi.) = 1 pint, pt.
2 pints = 1 quart, qt.
4 quarts = 1 gallon, gal.
1 gal. = 4 qt. = 8 pt. = 32 gi.

A gallon contains 231 cubic inches.
The standard barrel is 31½ gal., and the hogshead 63 gal

129. Apothecaries' Fluid Measure is used in mixing medicines in liquid form.

TABLE.

60 minims (♏) = 1 fluid dram, f. ʒ.
8 fluid drams = 1 fluid ounce, f. ℨ.
16 fluid ounces = 1 pint (O).

133. Dry Measure is used in measuring roots grain, vegetables, etc.

TABLE.

2 pints = 1 quart, qt.
8 quarts = 1 peck, pk.
4 pecks = 1 bushel, bu.
1 bu. = 4 pk. = 32 qt. = 64 pt.

The bushel contains 2150.42 cubic inches.

134. Avoirdupois Weight is used in weighing all common articles; as, coal, groceries, hay, etc.

TABLE.

16 ounces = 1 pound, lb.
100 pounds = { 1 hundred-weight, cwt.
              { or cental, ctl.
20 cwt., or 2000 lb. = 1 ton, T.
1 T. = 20 cwt. = 2000 lb. = 32000 oz.

The Long Ton of 2240 pounds is used at the U. S. Custom-House and in weighing coal at the mines.

The ounce is considered as 16 drams.

The Avoirdupois pound contains 7000 grains.

A hundred-weight is sometimes called a Cental.

135. Troy Weight is used in weighing gold, silver, and jewels.

TABLE.

24 grains (gr.)   = 1 pennyweight, pwt.
20 pennyweights = 1 ounce,   oz.
12 ounces   = 1 pound,   lb.
1 lb. = 12 oz. = 240 pwt. = 5760 grains.

136. Apothecaries' Weight is used by druggists and physicians in weighing medicines that are not liquid.

TABLE.

20 grains (gr.) = 1 scruple, sc. or ℈.
3 scruples   = 1 dram,   dr. or ʒ.
8 drams   = 1 ounce,   oz. or ℥.
12 ounces   = 1 pound,   lb. or ℔.
1 lb. = 12 oz. = 96 dr. = 288 sc. = 5760 gr.

Dry medicines are bought and sold in large quantities by avoirdupois weight.

Comparison of Weights.

1 lb. Avoirdupois   = 7000 gr.
1 oz. Avoirdupois   = 437½ gr.
1 lb. Troy or Apothecary   = 5760 gr.
1 oz. Troy or Apothecary   = 480 gr.

137. Measure of Time.

TABLE.

60 seconds (sec.) = 1 minute, min.
60 minutes   = 1 hour,   hr.
24 hours   = 1 day,   da.
7 days   = 1 week,   wk.
365 days   = 1 year,   yr.
366 days   = 1 leap year.
100 years   = 1 century.

The Civil Day begins and ends at midnight.

The exact time in which the earth makes one revolution of the sun is 365 da. 5 hr. 48 min. 49.7 sec., or 365¼ days, nearly. For convenience the common year is regarded as 365 days; the fraction being disregarded until it amounts to a full day, which is in four years, nearly. Accordingly every fourth year contains 366 days. This day is added to the shortest month, February, and the year in which it is added is called Leap Year.

But 365¼ days is a little more than the exact year, and we have added a little too much, when we have added 1 day to every fourth year, therefore only every fourth centennial year is considered as leap year. This nearly corrects the excess, so that the error is less than 1 day in about 3600 years.

Every year divisible by 4, and every centennial year divisible by 400, is a Leap Year.

Circular or Angular Measure.

138. A **Circle** is a plane figure bounded by a curved line, every point of which is equally distant from the centre.

139. The bounding line of a circle is the **Circumference.**

140. Any part of a circumference is an **Arc.**

A to C and B to D are arcs of a circle.

141. A straight line through the centre of a circle terminating at the circumference is the Diameter.

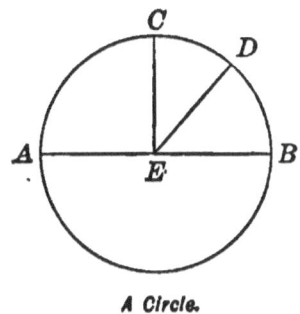

A Circle.

142. A straight line from the centre to the circumference is the radius; as, E to D, or E to C.

143. The circumference of every circle is divided into 360 equal parts called **Degrees,** each degree into 60 parts called **Minutes,** and each minute into 60 parts called **Seconds.**

144. An **Angle** is the difference in direction between two straight lines. The point of meeting is the **Vertex.** The vertex is at the centre of a circle, and the angle is measured in degrees by the arc between its sides. Thus B D is the measure of the angle B E D.

TABLE OF CIRCULAR MEASURE.

60 seconds ($''$) = 1 minute, $'$
60 minutes   = 1 degree, $°$
360 degrees   = 1 circumference, Cir.

NOTE.—An arc of 90 degrees or ¼ of a circumference is called a quadrant. A degree upon a great circle of the earth is 69.16 statute miles, or 60 geographical miles. A sign is an arc of 30 degrees.

146. Federal Money is the currency of the United States.

TABLE.

10 mills = 1 cent, ct.   10 dimes   = 1 dollar, $
10 cents = 1 dime, d.   10 dollars = 1 eagle, E.

The gold coins of the United States are the double-eagle, eagle, half-eagle, quarter-eagle, and one-dollar piece.

The silver coins are the dollar, half-dollar, quarter-dollar, and the ten-cent piece.

The five-cent piece is nickel, and the one-cent piece bronze.

147. English or Sterling Money.

TABLE.

4 farthings = 1 penny,   d.
12 pence   = 1 shilling, s.
20 shillings = 1 pound,   £, or 1 sovereign.

The coin which represents the Pound Sterling is the Sovereign, equal in value to $4.8665.

COMPOUND NUMBERS. 49

148. Counting.

TABLE.

12 things = 1 dozen,   doz.
12 dozen = 1 gross,   gr.
12 gross = 1 great gross, G. gr.

149. Paper.

TABLE.

24 sheets = 1 quire.   2 reams = 1 bundle.
20 quires = 1 ream.   5 bundles = 1 bale.

REDUCTION DESCENDING.

150. 1. Reduce 5 lb. 6 oz. 12 pwt. 6 gr. to grains.

```
5 lb. 6 oz. 12 pwt. 6 gr.
  12
  60
   6
  66 oz.
  20
 1320
   12
 1332 pwt.
   24
 5328
 2664
31968
    6
31974 gr.
```

Since there are 12 oz. in 1 lb., in 5 lb. there are 5 times 12 oz. = 60 oz. (add 6 oz.) = 66 oz.

Since there are 20 pwt. in 1 oz., in 66 oz. there are 66 times 20 pwt. = 1320 pwt. (add 12 pwt.) = 1332 pwt.

Since there are 24 gr. in 1 pwt., in 1332 pwt. there are 1332 times 24 gr. = 31968 gr. (add 6 gr.) = 31974 gr.

Reduce to lower denominations:

2. 17 yd. 2 ft. 9 in. to inches.
3. 46 rd. 4 yd. 2 ft. to feet.
4. 3 mi. 75 rd. 4 ft. to inches.
5. 16 A. 140 sq. rd. 26 sq. yd. to square yards.
6. 4 A. 15 sq. rd. 4 sq. ft. to square inches.
7. 50 ch. 45 li. to links.
8. 16 cu. yd. 25 cu. ft. 900 cu. in. to cubic inches.
9. 8 cd. 12 cu. ft. to cubic feet.
10. 15 gal. 3 qt. 1 pt. to pints.

SENIOR ARITHMETIC.

11. 4 O. 6 f. ʒ 3 f. ʒ 25 ℳ to minims.
12. 7 bu. 3 pk. 5 qt. 1 pt. to pints.
13. 16½ bu. to quarts.
14. 25 lb. 5 oz. 16 pwt. 10 gr. to grains.
15. 2 T. 6 cwt. 10 lb. 14 oz. to ounces.
16. 16 ℔. 5 ʒ 4 ʒ 2 ⴱ 11 gr. to grains.
17. 28° 14′ 18″ to seconds.
18. £18 15s. 8d. 3 far. to farthings.
19. 27 da. 18 h. 49 min. to seconds.
20. 3 wk. 48 min. 52 sec. to seconds.
21. How many quires in a bundle of paper?
22. How many pints in a cask of molasses holding 84 gallons?
23. How many articles in 7 G. gr. 5 gr.?
24. How many hours in 10 years, allowing for two leap years?
25. How many inches in 4½ rods?
26. What is the cost of 10 miles of telephone wire at 28 cents a pound, if a pound measures 75 ft.?
27. Find the number of square inches in a square yard; square feet in a square chain; cubic inches in a cubic yard.
28. How many hours in the month of February, 1896?
29. How many cubic inches in 5 gallons.
30. How many square yards in 4 sq. miles?
31. How many square feet in 2½ acres?
32. How many ounces in 3 lb. of silver? 3 lb. of iron?
33. If I buy 3 bu. of nuts at $4 a bushel, and sell them at 5¢ a pint, how much shall I gain?
34. How many ounces in a long ton?

COMPOUND NUMBERS. 51

35. At $12 a ton, what will ¾ of a ton of hay cost?

36. In 1800 years how many centuries?

37. If you can count sixty a minute, how long will it take to count 180000.

38. Through how many degrees does the hour-hand of a clock pass in 6 hours?

39. Through how many degrees does the minute-hand pass in 6 hours?

40. What will 3 reams of paper cost at 40¢ a quire?

41. Reduce 3 mi. 4 fur. 20 rd. 5 yd. 2 ft. 8 in. to inches?

42. Reduce 6 mi. 240 rd. to feet.

43. Reduce 3 A. 8 sq. rd. 5 sq. yd. 3 sq. ft. to sq. inches.

44. Reduce 16 cu. yd. 9 cu. ft. 3 cu. in. to cu. inches.

45. Reduce 58 cd. to cu. feet.

46. Reduce 2 T. 3 ctl. 16 lb. to ounces.

47. Reduce 3 lb. 9 oz. 15 pwt. 12 gr. to grains.

48. Reduce 60 gal. 3 qt. 3 gi. to gills.

49. How many sheets in 5 bales of paper?

50. Reduce 3 wk. 6 da. 5 hr. to minutes.

REDUCTION ASCENDING.

151. 1. Reduce 1306 gills to higher denominations.

4 / 1306 gi.
2 / 326 pt. + 2 gi.
4 / 163 qt.
40 gal. + 3 qt.
40 gal. 3 qt. 2 gi.

Since in 1 pt. there are 4 gi., in 1306 gi. there are as many pints as 4 gi. is contained times in 1306 gi., or 326 pt. and 2 gi. remainder.

Since in 1 qt. there are 2 pt., in 326 pt. there are as many quarts as 2 pt. is contained times in 326 pt., or 163 qt.

Since in 1 gal. there are 4 qt., in 163 qt. there are as many gallons as 4 qt. is contained times in 163 qt., or 40 gal., and 3 qt. remainder.

Therefore, in 1306 gills there are 40 gal. 3 qt. 0 pt. 2 gi.

2. How many rods in 334 yd.?

5½ yd.   334 yd.
  2   2
11 half yd.) 668 half yd.
              60 rd. + 8 half yd.

Since in 1 rd. there are 5½ yd., in 334 yd. there are as many rods as 5½ yd. is contained times in 334 yd., or 60 rd., and 4 yd.

remainder. 334 yd. ÷ 5½ yd. = 668 half yd. ÷ 11 half yd. 8 half yd. = 4 yd. 334 yd. = 60 rd. 4 yd.

3. Reduce 225932 in. to miles, etc.
4. How many miles and rods are there in 35640 ft.?
5. Reduce 19922544 sq. in. to higher denominations.
6. Reduce 762051 cu. in. to cu. yards, etc.
7. How many cords in 7424 cu. ft.
8. Reduce 69056 oz. to tons, etc.
9. Reduce 21076 gr. to higher denominations.
10. Reduce 1947 gi. to gallons, etc.
11. How many bales in 24000 sheets of paper?
12. Reduce 39180 min. to weeks, etc.
13. Reduce 5762 far. to higher denominations.
14. Reduce 84623″ to higher denominations.
15. Reduce 62341 M. to higher denominations.
16. How many chains, etc., in 13025 li.?
17. How many bushels, etc., in 35842 pints?
18. How many pounds, etc., (Troy) in 32563 gr.?
19. Reduce 39632 gr. to lb., etc. (Apoth.).
20. How many tons, etc., in 35682 lb.?
21. A box contains 75832 pens. How many Gr. gross, etc., in the box?
22. Change 1384 dry pints to higher denominations.
23. In 139843 sq. in. how many miles, rods, etc.?
24. How many cords of wood in 3692 cu. feet?

COMPOUND NUMBERS.

REVIEW PROBLEMS.

152. 1. Bought 2 gal. 8 oz. of fluid extract at 20¢ an ounce, and sold it at 15¢ an ounce. What was lost?

2. How many minims are there in 10 fluid ounces (f. ℥), 7 fluid drachms (f. ʒ)?

3. Find the difference in value between 4 gal. of ammonia water at 10 cents a pint and 8 ounces of cinnamon water at 5 cents an ounce.

4. The pendulum of a certain clock beats seconds. How many times will it tick in 1 day, 9 hours, 25 minutes?

5. How many degrees in 3492.58 statute miles, measured on the equator, a degree being equal to 69.16 statute miles?

6. How many degrees of longitude will a steamship pass through, sailing due west on the equator, at the rate of 15 knots an hour for 5 days?

NOTE. — A knot = 1 geographic mile or minute.

7. Find cost of each of following:

(a) 5 gallons, 3 qt. 1 pt. of molasses at 20¢ a gallon;

(b) 10 acres, 50 sq. rd. of land at $50 an A.

8. What will it cost to build 112 rd. 3 yd. of fence at 48¢ a yard?

9. If a man steps 2½ ft. at each step, how many miles will he travel in stepping 4820 times?

10. If 17 ft. is ⅝ of the height of a tree, how high is the tree?

11. Reduce 6⅔ mi. 317 rd. 4 yd. 2 ft. to feet.

12. Change 16571 ft. to miles.

13. At $3.20 a bu. how many quarts of nuts can be bought for $4.80?

14. How many pint bottles of camphor may be filled from 96 fluid ounces (f. ℥)?

15. Find the cost of the following: 4 oz. iodine at 10¢, 8 oz. spts. camphor at 5¢, 10 oz. aqua ammonia at 10¢, 14 oz. cinnamon water at 5¢.

16. Reduce 4 bu. 3 pk. 7 qt. 1 pt. to pints.

17. How many quart boxes will hold 2 bu. 3 pk. 5 qt. of berries?

18. If 4 bu. of berries are bought for $.70 per bushel and sold for $.05 per quart, what is the gain or loss?

19. Both sides of a railroad track are fenced with wire for 40 yards. What is the cost of the fence at 4¢ a foot?

20. What will 8 lb. 6 oz. of sugar cost at 8¢ a pound?

21. When pens are bought at 75¢ a gross, and sold at 2 for 3¢, what is the gain?

22. If a man can walk 10 miles in 2 hours, how far can he walk in 6 hours? in 30 minutes? in 50 minutes?

23. What will $\frac{1}{2}$ bu. berries bring at 8¢ a quart?

24. A silver chain weighs 18 pwt.; what is its value, when silver is worth $.65 an ounce?

25. What will 24 qt. of milk cost at 20¢ a gallon?

26. If I buy peanuts at 5¢ a quart, and retail them so as to gain $6.40 on 4 bushels, what do I sell them for?

27. At 4 pens for 3 cents what will 1 great gross cost?

28. How many table-forks, each weighing $2\frac{1}{2}$ oz., can be made from 4 lb. 4 oz. 10 pwt. of silver.

29. In $\frac{3}{4}$ of a gallon how many pints.

30. How many rods of fence will enclose a mile square of land?

31. What is the cost of 1 yd. and 27 inches of fringe at 60 cents a yard.

32. How many rods of fence are required to enclose a lot that is 20 rods wide and three times as long?

COMPOUND NUMBERS. 55

33. Required the distance around a room that is 13 feet long and 15 feet wide.

34. A shoe-box is 4 in. deep, 6 in. wide, and 12 in. long. How much twine will it take to wind twice around the box each way to hold on the cover, allowing 6 inches for tying.

35. I have a lawn that is 30 ft. by 70 ft., and wish to lay a board walk around it that is 3 ft. 6 in. in width. What is the distance around the walk, outside measurement?

153. A **Denominate Fraction** is a fraction having a denomination.

154. To reduce Denominate Fractions to Integers of Lower Denominations.

1. Reduce $\frac{5}{7}$ of a mile to rods, yards, feet, etc.

SOLUTION. — $\frac{5}{7}$ of 320 rd. = $\frac{1600}{7}$ rd. = $228\frac{4}{7}$ rd.

$\frac{4}{7}$ of $5\frac{1}{2}$ yd. = $\frac{22}{7}$ yd. = $3\frac{1}{7}$ yd.

$\frac{1}{7}$ of 3 ft. = $0\frac{3}{7}$ ft.

$\frac{3}{7}$ of 12 in. = $\frac{36}{7}$ in. = $5\frac{1}{7}$ in.

$\frac{5}{7}$ of a mile = 228 rd., 3 yd., 0 ft., $5\frac{1}{7}$ in.

NOTE. — The same process applies to denominate decimals.

2. Reduce .87 bu. to pecks, quarts, etc.

.87 of 4 pk. = 3.48 pk.   87 bu.
                                    4
.48 of 8 qt. = 3.84 qt.   3.48
                                    8
.84 of 2 pt. = 1.68 pt.   3.84
                                    2
.87 bu. = 3 pk., 3 qt., 1.68 pt.   1.68

Rule. — *Change the given fraction (or decimal) to the next lower denomination. Treat the fractional (or decimal) part of the product in the same way, and so proceed to the required denomination.*

Reduce to integers of lower denominations.

3. $\frac{2}{3}$ of a mile.
4. $\frac{6}{8}$ of an acre.
5. $\frac{4}{7}$ of a pound (Troy).
6. $\frac{3}{7}$ of a ton.
7. $\frac{5}{8}$ of a gallon.
8. $\frac{5}{8}$ of a mile.

9. .375 of a month.
10. .3125 of a gallon.
11. .4267 of an acre.
12. .2364 of a ton.
13. .363 of a sign.
14. .51625 of a mile.

15. Reduce $1\frac{7}{8}$ mi. to lower denominations.
16. Change $\frac{3}{7}$ of a year to months and days.
17. In $\frac{7}{12}$ gal. how many qt. and pt.?
18. Reduce $\frac{25}{121}$ lb. to oz. and dr.
19. $\frac{9}{14}$ acres are equal to how many sq. rods, etc.?
20. Reduce $\frac{3}{8}$ bu. to integers of lower denominations.
21. What is the value of $\frac{5}{9}$ of $\frac{5}{8}$ of a hhd. in integers of lower denominations?
22. What is the value of $\frac{7}{13}$ of an acre in integers of lower denominations?
23. Reduce £$\frac{1}{3}$ to integers of lower denominations.
24. What is the value of $\frac{1}{8}$ of $1\frac{1}{3}$ of a mile?

155. To reduce denominate numbers to Fractions of Higher Denominations.

1. Reduce 2 qt. 1 pt. 2 gi. to the fraction of a gallon.

SOLUTION.— 2 gi. $\div 4 = \frac{2}{4}$ pt. $= \frac{1}{2}$ pt.
$1\frac{1}{2}$ pt. $= \frac{3}{2}$ pt. $\div 2 = \frac{3}{4}$ qt.
$2\frac{3}{4}$ qt. $= 1\frac{1}{4}$ qt. $\div 4 = \frac{11}{16}$ gal. *Ans.*

2. Reduce 2 qt. 1 pt. 2 gi. to the decimal of a gallon.

Rule.— Change the number of the lowest denomination to a fraction (or decimal) of the next higher, write this fraction (or decimal) as a part of the number of that higher denomination, and reduce

4 / 2 gi.
2 / 1.5 pt.
4 / 2.75 qt.
.6875 gal.

COMPOUND NUMBERS. 57

this number in like manner, and so proceed to the required denomination.

3. Reduce 213 rd. 1 yd. 2 ft. 6 in. to a fraction of a mile.

4. What fraction of an acre is 3 R. 13 sq. rd. 10 sq. yd. 108 sq. in. ?

5. What part of a year is 273 da. 18 hr. ?

6. Reduce to a fraction of a pound 8 oz. 11 pwt. 10⅔ gr.

7. What part of a ton is 857 lb. 2⅔ oz. ?

8. Change 3 fur. 19 rd. 5 yd. 1 ft. 4.7328 in. to the decimal of a mile.

9. Reduce 1 da. 14 hr. 24 min. to the decimal of a month.

10. What decimal of a gallon is 1 qt. 2 gi. ?

11. Reduce 68 sq. rd. 8 sq. yd. 2 sq. ft. 7.488 sq. in. to the decimal of an acre.

12. What decimal of a pound Troy is 6 oz. 3 pwt. 21.6 gr. ?

13. Reduce 131 da. 18 hr. 21 min. 36 sec. to the decimal of a year.

14. Reduce 2 qt. 1⅜ gi. to the fraction of a gallon.

15. What fraction of a mile is 71 rd. 1 ft. 10 in. ?

16. Reduce 12 da. 34 min. 14½ sec. to the fraction of a month.

17. What decimal of a ton is 4 cwt. 72 lb. 128 oz. ?

18. Reduce 48 cu. ft. 1636.7616 cu. in. to the decimal of a cord.

19. What decimal of a circle is 10° 53′ 24″ ?

20. Reduce 4 fur. 5 rd. 1 yd. 3.6 in. to the decimal of a mile.

21. Reduce 6 pwt. to a fraction of a pound.
22. 3 qt. 1 pt. 2 gills are what part of a peck?
23. Change 6 rd. 4 yd. 1 ft. to the fraction of a mile.
24. What part of a cord of wood are 8 cu. ft.?
25. Reduce 5 gross 7 doz. to the fraction of a score.

To find what part one denominate number is of another.

1. What part of 2 gal. 1 qt. 1 pt. is 3 qt. 1 pt. 1 gi.?

$$3 \text{ qt. } 1 \text{ pt. } 1 \text{ gi.} = 29 \text{ gi.}$$
$$2 \text{ gal. } 1 \text{ qt. } 1 \text{ pt.} = 76 \text{ gi.}$$

The question now is, 29 gi. is what part of 76 gi.?

29 gi. is $\frac{29}{76}$ of 76 gi. *Ans.*

NOTE. — To find the decimal part, divide numerator by denominator.

2. What part of 5 lb. 9 oz. 3 pwt. is 2 lb. 8 oz. 6 pwt. 10 gr.?
3. What part of 3 mi. 24 rd. 5 yd. is 2 mi. 34 rd. 4 yd.?
4. What part of 3 da. 5 hr. 22 min. is 1 da. 10 hr. 3 min. 12 sec.?
5. What decimal of 3 gal. 2 qt. 1 pt. is 2 gal. 2 qt. 2 pt.?
6. What decimal of 4 T. 5 cwt. 10 lb. is 2 T. 6 cwt. 13 lb.?
7. What part of a rod is 4 yd. 2 ft. 7 in.?
8. What part of 6 rods is $\frac{5}{6}$ of 7 feet?
9. What part of $3\frac{3}{4}$ mi. is 160 rd. 5 yd.?
10. $\frac{1}{2}$ pint is what part of a bushel?
11. What decimal of 8 bu. 3 pk. 4 qt. is 4 bu. 1 pk. 5 qt.

ADDITION OF COMPOUND NUMBERS.

156. 1. Add 14 lb. 5 oz. 17 pwt. 12 gr., 18 lb. 10 oz. 14 gr., 6 lb. 4 oz. 8 pwt. 16 gr.

COMPOUND NUMBERS.

lb.	oz.	pwt.	gr.
14	5	17	12
18	10	0	14
6	4	8	16
39	8	6	18

SOLUTION.—The sum of the grains = 42 gr. = 1 pwt. 18 gr. We place the 18 gr. under the column of grains, and add the 1 pwt. to the column of pennyweights. Add the other columns in like manner.

2.
rd.	yd.	ft.
17	4	1
12	4	2
6	5	2½
8	3	2
46	1½	1½
		1½ = ½ yd.
46	2	0

3.
rd.	ft.	in.
6	12	6
4	14	11
17	15	9
6	12	8
36	5½	10
		6 = ½ ft.
36	6	4

4.
tons.	cwt.	lb.	oz.
14	13	65	15
13	17	88	11
46	16	86	13
14	15	57	6
11	17	85	15

7.
yr.	da.	hr.	min.	sec.
18	345	13	37	15
87	169	12	16	28
316	144	20	53	18
13	360	21	57	15

5.
deg.	min.	sec.
29	59	59
15	45	42
18	11	40
13	19	17

8.
bu.	pk.	qt.	pt.
40	2	6	1
89	1	3	0
75	2	1	1
69	2	3	0
49	1	2	1
65	3	1	1

6.
sq. yd.	sq. ft.	sq. in.
45	8	113
45	3	112
75	8	139
49	0	115
589	8	90

9.
cd.	cd. ft.	cu. ft.
5	7	0
2	2	12
0	6	15
7⅜	0	0
3	0	2

60 SENIOR ARITHMETIC.

10. Find the sum of 3 T. 15 cwt. 25 lb. 9 oz., 4 T. 17 cwt. 30 lb. 10 oz., 6 T. 18 cwt. 15 lb. 12 oz., 2 T. 12 cwt. 20 lb. 16 oz.

11. Find the sum of 7 hr. 30 min. 45 sec., 12 hr. 25 min. 30 sec., 20 hr. 15 min. 33 sec., 10 hr. 27 min. 46 sec.

12. Add 10 mi. 101 rd. 3 yd. 2 ft. 11 in., 16 mi. 4 yd. 6 in., 3 mi. 560 rd. 3 ft., 175 rd. 4 ft. 7 in.

13. Add 3 A. 50 sq. rd. 25 sq. yd. 10 sq. ft. 102 sq. in., 5 A. 110 sq. rd. 30 sq. yd. 8 sq. ft. 34 sq. in., 6 A. 75 sq. rd. 14 sq. yd., 7 sq. ft., 82 sq. in., 7 A. 215 sq. rd., 17 sq. yd. 16 sq. ft. 53 sq. in.

14. Find the sum of 18 cd. 6 cd. ft. 12 cu. ft., 19 cd. 4 cd. ft. 6 cu. ft., 24 cd. 2 cd. ft. 1 cu. ft.

15. Find the sum of 18 T. 18 lb. 12 oz., 16 cwt. 21 lb., 14 cwt. 75 lb. 10 oz.

16. What is the entire length of a railway consisting of 5 different lines measuring respectively 160 mi. 185 rd. 2 yd., 97 mi. 63 rd. 4 yd., 126 mi. 272 rd. 3 yd., 67 mi. 199 rd. 5 yd., and 48 mi. 266 rd. 5 yd. ?

17. A merchant sold 48 gal. 3 qt. 1 pt. of coal oil and had 15 gal. 1 qt. 1 pt. left. What quantity had he at first ?

18. A starts from a point in Lat. 21° 25′ 35″ N. and travels north 24° 36′ 45″. At what latitude does he arrive ?

19. Find the difference in longitude between a point 46° 15′ 30″ E. and a point 21° 18′ 16″ W.

NOTE. — When one place is in east and the other in west longitude, their difference in longitude is the sum of their longitudes.

20. Charles walks 5 mi. 15 rd. 2 ft. north of the schoolhouse, and James 6 mi. 28 rd. 5 yd. south. How far are they apart ?

21. Find the sum of $\tfrac{2}{3}$ mi. 35 rd. $4\tfrac{3}{8}$ rd.

COMPOUND NUMBERS. 61

Note. — Reduce each to integers of lower denominations, then add.

22. Add $\frac{2}{3}$ bu. 17$\frac{1}{2}$ pk. 4$\frac{3}{4}$ pt., 6 bu. 3$\frac{2}{3}$ pk. 2 qt., $\frac{1}{4}$ bu. $\frac{1}{2}$ pk. 5 qt.

23. What is the sum of $\frac{3}{5}$ T. $\frac{5}{8}$ cwt. and $\frac{5}{8}$ lb. ?

SUBTRACTION OF COMPOUND NUMBERS.

	lb.	oz.	pwt.	gr.
157. 1. From	6	2	14	15
Take	4	10	18	12
	1	3	16	3

Solution. — 15 gr. — 12 gr. = 3 gr. As we cannot take 18 pwt. from 14 pwt., we take 1 oz., which equals 20 pwt., and add to the 14 pwt. = 34 pwt.; 34 pwt. — 18 pwt. = 16 pwt. We have taken 1 oz. from the 2 oz., leaving 1 oz.; but as we cannot take 10 oz. from 1 oz., we take 1 lb = 12 oz., and add it to 1 oz. = 13 oz., from which take 10 oz. = 3 oz. Since we took 1 of the 6 lb., we have 5 left; from which take 4 lb. = 1 lb.

	2.				3.		
	A.	sq. rd.	sq. ft.		hr.	min.	sec.
From	10	50	7		5	54	30
Take	4	106	5		1	17	50

	4.				5.				
	gal.	qt.	pt.	gi.	A.	R.	sq. rd.	sq. yd.	sq. ft.
From	39	2	2	1	5	1	39	15	7
Take	16	2	3	3	2	2	26	21	8

	6.				7.			
	da.	hr.	min.	sec.	T.	cwt.	lb.	oz.
	200	17	54	36	20	15	75	10
	135	20	24	48	5	16	25	12

7. From 260 mi. take 23 mi. 7 fur. 25 rd. 5 yd. 2 ft. 10 in.

8. A man having ½ an acre of ground, sold 25 sq. rd. 11 sq. yd. 8 sq. ft. to one man, and 50 sq. rd. 9 sq. yd. 4 sq. ft. to another. How much land had he left?

9. From 12 cwt. subtract 9 cwt. 14 lb. 12 oz.

10. From a hogshead of molasses, 25 gal. 3 qt. 2 pt. were drawn at one time, and at another time 10 gal. 1 pt. How many gallons remained?

11. From 2 bu. 3 pk., 1 bu. 2 pk. 6 qt. were sold. How much remained?

12. An apothecary bought 2 lb. of quinine, and sold 1 lb. 3 oz. 5 dr. 2 sc. 11 gr. How much had he left?

13. What is the difference in longitude between New York (74° 0′ 3″ W.) and San Francisco (122° 25′ 40″ W.).

Note.—When both places are in east or in west longitude, their difference in longitude is found by subtraction.

14. Charles walks 25 mi. 4 rd. 2 yd. south of the school, and Henry 16 mi. 160 rd. 3 yd. in the same direction. How far are they apart?

15. From ⅔ of a mile take 16¼ rods.

Note.—Change both to integers of lower denominations, then subtract.

16. Rome is in longitude 12° 28′ 40″ E., and Paris in long. 2° 20′ 14″ E. What is their difference in longitude?

17. From ⅚ of a pound Avoir. take 3⅔ oz.

18. From 16⅔ bu. take 7½ pk.

19. From .325 T. take 6.54¾ cwt.

20. From .7 of a rod take 4 yd. 2 ft. 8 in.

21. From 22 da. 16 hr. 20 min. take 2¼ weeks.

COMPOUND NUMBERS. 63

DIFFERENCE BETWEEN DATES.

158. 1. Find the time from Jan. 25, 1842, to July 4, 1896.

1896	7	4
1842	1	25
54 yr.	5 mo.	9 da.

It is customary to consider 30 days to a month. July 4, 1896, is the 1896th yr. 7th mo. 4th da., and Jan. 25, 1842, is the 1842d yr. 1st mo. 25th da. Subtract, taking 30 da. for a month.

2. What is the exact number of days between Dec. 16, 1895, and March 12, 1896 ?

Dec. 15
Jan. 31
Feb. 29
Mar. 12
——
87 days.

Do not count the first day mentioned. There are 15 days in December, after the 16th. January has 31 days, February 29 (leap year), and 12 days in March ; making 87 days. *Ans.*

3. How much time elapsed from the landing of the Pilgrims, Dec. 11, 1620, to the Declaration of Independence, July 4, 1776 ?

4. How much time elapsed from the beginning of the Civil War, April 14, 1861, to the close of the war April 9, 1865 ?

5. Washington was born Feb. 22, 1732, and died Dec. 14, 1799. How long did he live ?

6. Washington was first inaugurated April 30, 1789. How long ago was his inauguration ?

7. How much time will have elapsed since Columbus discovered America, Oct. 12, 1492, to your next birthday ?

8. Mr. Smith gave a note dated Feb. 25, 1896, and paid it July 12, 1896. Find the exact number of days between its date and time of payment.

9. A carpenter earning $2.50 per day, commenced Wednesday morning, April 1, 1896, and continued working every week day until June 6. How much did he earn ?

SENIOR ARITHMETIC.

10. Fred was born Dec. 20, 1875; how old is he now?

11. How much time has elapsed since George Washington was $15\frac{1}{2}$ years old?

12. Gen. Grant was born April 27, 1822. How old would he be if he were alive to-day?

13. How long since Lee surrendered to Gen. Grant?

14. Find the exact number of days between Jan. 10, 1896, and May 5, 1896.

15. When can a boy who was born May 5, 1882, celebrate his 25th birthday?

16. John goes to bed at 9.15 P.M. and gets up at 7.10 A.M. How many minutes does he spend in bed?

MULTIPLICATION OF COMPOUND NUMBERS

159. 1. Multiply 4 yd. 2 ft. 8 in. by 8.

yd.	ft.	in.
4	2	8
		8
39	0	4

8 times 8 in. = 64 in. = 5 ft. 4 in. Place the 4 in. under the inches column, and reserve the 5 ft., to be added to the product of 2 ft. by 8, which equals 16 ft. (add 5 ft.) = 21 ft. 21 ft. ÷ 3 = 7 yd., with no remainder. Add 7 yd. to the product of 4 yd. by 8 = 32 yd. (add 7 yd.) = 39 yd.

2. gal.	qt.	pt.	gi.
31	3	2	3
			5

bu.	pk.	qt.	pt.
12	3	2	1
			8

3. If a man travel at the rate of 60 mi. 240 rd. 16 ft. in one day, how far will he travel in ten days?

4. A man owns 6 farms, each containing 75 A. 49 sq. rd. 25 sq. yd. of land. How much land in all the farms?

5. If 6 loads of hay weigh 6 T. 18 cwt. 75 lb., how much will 48 loads weigh?

NOTE. — 48 loads will weigh 8 times as much as 6 loads.

COMPOUND NUMBERS.

6. If 12 spoons weigh 3 lb. 8 oz. 15 pwt., how much will one gross of spoons weigh?

7. How much oil will 7 barrels hold if each barrel contains 35 gal. 2 qt.?

8. What is the value at $4 per cord of 10 piles of wood, each containing 5 cd. 5 cd. ft. 12 cu. ft.?

9. What is the weight of 15 packages, each weighing 1 lb. 4 oz. (Avoir.)?

10. If a bicyclist travels 75 mi. 140 rd. in one day, how far can he travel in ten days?

11. In a watch-chain there are 2 oz. 12 pwt. 15 gr. of gold. How much gold is required for 25 such chains?

12. A farmer has six bins, each containing 60 bu. 2 pk. of wheat. How much wheat has he?

13. If a train is run for 8 hours at the average rate of 50 mi. 30 rd. 10 ft. per hour, how great is the distance covered?

14. It takes John Smith 5 hr. 20 min. 11 sec. to plough one acre of ground. At the same rate, how long will it take him to plough 4 acres?

15. 4 gal. 3 qt. 1 pt. × 11 = ?

16. 2 A. 40 sq. rd. 16 sq. yd. × 20 = ?

DIVISION OF COMPOUND NUMBERS.

160. 1. Divide 16 lb. 9 oz. 17 pwt. 8 gr. by 10.

```
      lb.  oz.  pwt.  gr.
10 / 16   9    17    8
     1    8    3    17 6/10
```

SOLUTION. — $\frac{1}{10}$ of 16 lb. = 1, and 6 lb. remaining. 6 lb. = 72 oz. 72 oz. + 9 oz. = 81 oz. $\frac{1}{10}$ of 81 oz. = 8 oz., with 1 oz. remaining, = 20 pwt., to which add 17 pwt., = 37 pwt. $\frac{1}{10}$ of 37 pwt. = 3 pwt., with 7 pwt. remaining, = 108 gr., to which add 8 gr.; and taking $\frac{1}{10}$ of the sum, we have $17\frac{6}{10}$ gr.

When the divisor is large, employ long division.

2. Find $\frac{1}{35}$ of 42 rd. 4 yd. 2 ft. 8 in.

```
   rd. yd. ft. in.
35/42  4   2   8 ( 1 rd.
   35
   ――
    7
    5½
    ――
    3½
   35
   ――
   38½
   + 4
35/ 42½ yd. ( 1 yd.
    35
    ――
     7½
     3
    ――
    22½ ft.
    + 2
35/ 24½ ft. ( 0 ft.
    12
    ――
    294
    + 8
35/ 302 ( 8$\frac{22}{35}$ in.
    280
    ――
     22

1 rd. 1 yd. 8$\frac{22}{35}$ in.
        Ans.
```

$\frac{1}{35}$ of 42 rd. = 1 rd.; remainder, 7 rd. = 38½ yd.; add 4 yd. = 42½ yd. $\frac{1}{35}$ of 42½ yd. = 1 yd.; remainder, 7½ yd., = 22½ ft.; add 2 ft. = 24½ ft. $\frac{1}{35}$ of 24½ ft. = 0 ft. 24½ ft. = 294 in.; add 8 in. = 302 in. $\frac{1}{35}$ of 302 in. = 8$\frac{22}{35}$ in.

NOTE. — When both dividend and divisor are compound, reduce them to the same denomination, and divide. The quotient will be abstract.

3. Divide 169 bu. 3 pk. 5 qt. by 7.

4. If a man travelled 607 mi. 169 rd. 11 ft. 6 in. in 10 days, what average distance did he travel in 1 day?

5. If one gross of spools weighs 44 lb. 9 oz., how much will one dozen weigh?

6. If one bottle holds 1 pt. 3 gi., how many dozen bottles will be required to hold 65 gal. 2 qt. 1 pt. ?

7. A man has 451 A. 138 sq. rd. 29 sq. yd. of land, which he wishes to divide equally among his six children. How much land will each child receive?

8. If 12 persons share equally in the contents of a bin containing 20 bu. 2 pk. 4 qt. of apples, what is the share of each?

9. If the entire area of 24 equal fields is 242 A. 20 sq. rd. 15 sq. yd., what is the size of each field?

COMPOUND NUMBERS. 67

10. A man walked 50 mi. 71 rd. 2 yd. in 15 hours. What was his rate per hour?

11. If it takes a man 12 hr. 35 min. 15 sec. to walk 45 miles, what is the average time taken for each mile? (Divide by the factors of 45.)

12. When $12 will buy 11 gal. 2 qt. 1 pt. of maple sirup, how much will $1 buy?

13. A man travelled 100 miles in 9 hours. What was the average rate per hour?

14. If a horse eats 12 qt. of oats per day, how long will 10 bu. 1 pk. 4 qt. last him?

15. If a package weighs 4 cwt. 15 lb., how many such packages will it take to weigh 3 T. 2 cwt. 25 lb.?

16. A man had 5 acres of land which he divided into 12 equal parts. How much land did each part contain?

17. Divide 102 T. 15 cwt. 27 lb. 9 oz. by 8.

18. Divide 16 bu. 3 pk. 6 qt. by 2 bu. 1 pk.

19. I have 84 lb. 14 oz. of salt which I wish to put into packages of 2 lb. 6 oz. each. How many packages will there be?

20. If a horse eats 1 pk. 2 qt. of oats a day, how many days will 16 bu. 3 pk. 6 qt. last him?

21. How many sacks, each containing 2 bu. 3 pk. 2 qt., will be needed to hold 165 bu. 2 pk. of meal?

22. 16 cwt. 75 lb. 9 oz. of butter are to be put into jars each containing $9\frac{3}{4}$ lb. How many jars will be needed?

To multiply or divide a compound number by a fraction.

NOTE. — To multiply by a fraction, multiply by the numerator, and divide the product by the denominator.

To divide by a fraction, multiply by the denominator, and divide the product by the numerator.

23. How much is $\frac{6}{7}$ of 16 hr. 17 min. 14 sec. ?

24. How much is $\frac{7}{8}$ of 30 ℥ 5 ʒ 1 ∋ 8 gr. ?

25. Divide 120 cd. 50 cd. ft. 34 cu. in. by $\frac{3}{4}$.

26. How many times is $1\frac{1}{2}$ contained in 840 T. 15 cwt. 98 lb. 3 oz. ?

27. A man sold 4 bu. 3 pk. 2 qt. of potatoes, which was $\frac{4}{5}$ of what he raised; how much did he raise ?

28. A butcher sells 120 tons, 9 cwt. 75 lb. of beef every month. How much does he sell in $\frac{2}{3}$ of a month ?

29. If 6 bottles hold 5 gal. 2 qt. of milk, how much milk will 3 such bottles hold ?

30. A field contains 10 acres 12 sq. rd. of land, which is $\frac{4}{9}$ the size of the whole farm. Find the size of the farm.

31. A railroad track extends 144 miles, 40 rd. 3 yd. How far has a train of cars gone which has travelled $\frac{9}{16}$ of this distance ?

32. If a pipe discharges 25 gal. 3 qt. 1 pt. of water in 1 hr., how much will it discharge in $5\frac{3}{4}$ hr., if the water flows with the same velocity?

NOTE. — When the multiplier or divisor is a mixed number, reduce to an improper fraction, and proceed as above.

33. Divide 8 lb. 11 oz. 15 pwt. 18 gr. by $2\frac{2}{3}$.

34. If a railroad train runs 60 mi. 35 rd. 16 ft. in one hour, how far will it run in $12\frac{2}{5}$ hr. at the same rate of speed ?

35. Divide 14 bu. 3 pk. 6 qt. 1 pt. by $\frac{7}{8}$.

36. Divide 5 yr. 1 mo. 1 wk. 1 da. 1 hr. 1 min. 1 sec. by $3\frac{2}{5}$.

MISCELLANEOUS PROBLEMS.

161. 1. Name two numbers which multiplied together make 14.

2. Write three sets of factors for 24.

3. Find the prime factors of 2205.

4. $\dfrac{14 \times 6 \times 3 \times 2 \times 8}{5 \times 6 \times 2 \times 9 \times 24} = ?$

5. How many yards of silk, 24 inches wide, will it take to line a skirt containing six yards of cloth 28 inches wide?

6. Find the least common multiple of 2, 3, 4, 5, 6, 7, 8, 9.

7. Find the greatest common divisor of 285, 465.

8. Find the smallest number that will exactly contain 9, 15, 18, 20.

9. Find the length of the longest stick that will exactly measure the sides of a room 216 yd. by 111 yd.

10. What is the smallest sum of money with which you can buy pears at 75¢ a basket, peaches at 90¢, and grapes at 50¢, using the same amount of money for each kind?

11. How many times is 1 contained in ¼?

12. How many times is ¼ contained in 1?

13. A man bought a horse for $240, which is ⅔ of what he sold it for. What did it sell for?

14. A man bought a horse for $240, and sold it for ⅔ of what he paid for it. What did it sell for?

15. A pole stands ⅓ in the mud, ¼ in the water, and the remaining 10 feet are above the water. How long is the pole?

16. A man owns 4 farms containing $365\tfrac{1}{2}$, $375\tfrac{2}{3}$, $284\tfrac{3}{4}$. and $254\tfrac{2}{3}$ acres respectively. How many acres in all?

17. The man owning the above farms sells A $234\frac{4}{80}$ acres, and B $366\frac{3}{80}$ acres. How many acres has he left?

18. What is the value of $3\frac{8}{9} \times 1\frac{3}{16} \times \frac{7}{8} \times 14 \times 3\frac{9}{10} \times 1\frac{5}{8} \times 1\frac{6}{2} \times \frac{4}{8} \times 1\frac{2}{3} \times \frac{1}{2}$.

19. Find the least common multiple of 273, 462, 1785, and 399.

20. A man owns a farm containing 400 acres. He sells $\frac{1}{4}$ of the farm, and divides the remainder among his six children. How many acres does each child receive?

21. Find the sum of 3 bu. 6 pk. 2 qt. 1 pt., 3 pk. 1 qt. 1 pt., 7 bu. 3 pk., 4 bu. 7 qt. 1 pt., and 19 bu. 2 pk. 2 qt. 1 pt.

22. Find the sum of 4 lb. 6 oz. 21 pwt. 9 gr., 5 oz. 11 gr., 3 lb. 9 oz. 18 pwt., 11 oz. 17 pwt. 5 gr., 16 lb. 4 oz. 11 pwt., 18 lb. 17 gr., 21 lb. 15 pwt. 11 gr., 23 lb. 10 oz. 21 pwt. 23 gr.

23. What part of a mile is 214 rd. 2 yd. 2 ft. 3 in.?

24. Reduce 1 pk. 4 qt. $1\frac{3}{8}$ pt. to the fraction of a bushel.

25. What is the quotient of 184 bales, 4 bundles, 1 ream, 13 quires, 20 sheets, divided by $\frac{2}{3}$?

26. A farm is 60 ch. 25 l. long. How many rods long is it?

27. A surveyor measured my farm, and found that it is 80 ch. long and 60 ch. broad. How many acres does it contain?

28. In 133128 in. how many miles?

29. How many times will a wheel 12 ft. 3 in. in circumference turn round in going 15 mi. 20 rd. 12 ft. 2 in.?

30. How many steel rails 30 ft. long are needed in the construction of 7 mi. 305 rd. 7 ft. 6 in. of double-track railroad?

MISCELLANEOUS PROBLEMS.

31. What fraction of the year is contained in the months of July and August, 1896?

32. The greatest depth of the Atlantic telegraph cable is 2 mi. 250 rd. 5 yd. 1 ft. How many feet is it?

33. How many statute miles in 45° 22′ 30″, measured on the equator?

34. On an average when walking, Isaac steps 24 inches twice every second. How many minutes will it take him to walk $1\frac{1}{2}$ miles?

35. Reduce $\frac{9}{25}$ of an acre to square rods and decimals of a square rod.

36. A silversmith in making spoons uses 2 lb. 3 oz. 19 pwt. of silver in one day, 3 lb. 18 pwt. 20 grs. on the second day, and 11 oz. 19 pwt. 23 gr. on the third day. How much silver does he use altogether?

37. What decimal of a pound Troy are 4 oz. 14 pwt.?

38. Reduce 18 pwt. 164 gr. to the fraction of a pound.

39. Find the difference between $\frac{3}{4}$ lb. and 4 lb. 7.84 oz. Troy.

40. Find the cost of $2\frac{1}{2}$ doz. spoons, each weighing 9 oz. 8.76 pwt., at $.045 a pennyweight.

41. What decimal part of a grain is $\frac{1}{7240}$ of a pound?

42. What part of an acre is $\frac{3}{7}$ rd.?

43. Multiply 16 bu. 9 pk. 8 qt. by $\frac{3}{4}$.

44. What is the product of 20 bu. 9 oz. 11 pwt. 15. gr. multiplied by $\frac{1}{4}$?

MEASUREMENTS.

162. An **Angle** is the difference in direction of two lines.

163. A **Right Angle** is the angle of a square.

164. Anything that has length and breadth, but not thickness, is a **Surface**.

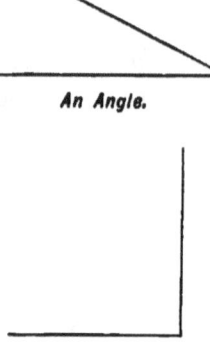

165. A surface that does not change its direction is a **Plane Surface**.

166. A figure having four straight sides and four right angles is a **Rectangle**.

167. A **Square** is a rectangle having equal sides.

168. A **Square Foot** is a square 1 foot long and 1 foot wide.

169. A **Square Yard** is a square 1 yd. long and 1 yd. wide.

170. The **Area** of a surface is the number of square units that it contains.

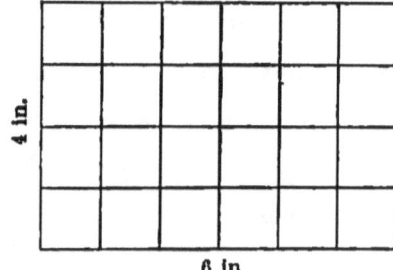

NOTE. — There are 6 sq. in. in a row, and in 4 rows there are 4 times 6 sq. in. = 24 sq. in.

The multiplier is abstract, and the unit of the product must be the same as the unit of the multiplicand.

MEASUREMENTS.

171. The length and breadth of a rectangle are called its **Dimensions**.

Length × Breadth = Area.
Area ÷ Length = Breadth.
Area ÷ Breadth = Length.

172. A figure having three straight sides and three angles is a **Triangle**.

The *Base* of a triangle is the line upon which it stands, and the *Altitude* is its height above the base, or the base extended. Thus, AC is the base, and BD the altitude, of the triangle shown below.

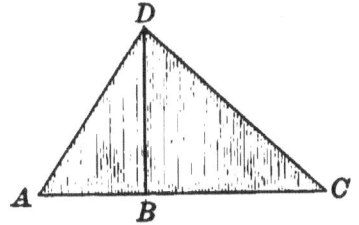 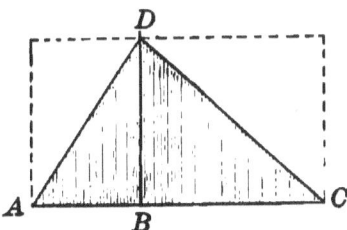

173. It is evident from the accompanying figure that *the area of a triangle is equal to one-half the area of a rectangle of the same base and altitude.*

Every circle may be regarded as composed of many equal triangles, the radius of the circle forming the altitudes, and the circumference forming the sum of the bases. Therefore, the area of a circle is equal to ½ the product of the circumference and radius.

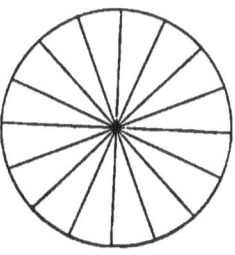

PRINCIPLE. — The circumference of a circle is 3.1416 times the diameter, or about $3\frac{1}{7}$.

174. Circumference ÷ 3.1416 = Diameter.
Diameter × 3.1416 = Circumference.

Oral.

Find the areas of rectangles as follows:

1. 10 ft. by 8 ft.
2. 16 ft. by 4½ ft.
3. 14 rods by 10 rd.
4. 50 ft. by 20 ft.
5. 6 ch. by 8 ch.
6. 9 yd. by 6 yd.

Find the other dimension.

7. Area 24 sq. ft., length 8 ft.
8. Area 72 sq. yd., length 8 yd.
9. Area 100 sq. in., breadth 5 in.
10. Length 16 yd., area 64 sq. yd
11. Breadth 4 ft., area 84 sq. ft.
12. Area 56 sq. ch., length 8 ch.

Find the areas of the following triangles:

13. Base 10 ft., alt. 12 ft.
14. Base 9 yd., alt. 6 yd.
15. Base 15 in., alt. 6 in.
16. Base 5 ft., alt. 10 ft.
17. Base 12 in., alt. 8 in.
18. Base 10 rd., alt. 5½ rd.

Find circumferences of circles having the following diameters:

NOTE. — Indicate the operation only.

19. 12 ft.
20. 18 ft.
21. 16 in.
22. 14 yd.
23. 16 rd.
24. 25 ch.
25. 62 yd.
26. 84 ft.

Find the diameters having the following circumferences:

NOTE. — Indicate only.

27. 78 ft.
28. 19 ft.
29. 316.14 rd.
30. 189.68 ch.

Written.

Find areas.

31. Circumference 37.6992 ft., radius 6 ft.
32. Circumference 47.124 ft., diameter 15 ft.

MEASUREMENTS. 75

33. Circumference 62.832 ft., diameter 20 ft.
34. Circumference 94.248 ft., diameter 30 ft.
35. Diameter 24 ft. 38. Diameter 160 ft.
36. Radius 16 ft. 39. Radius 62 ft.
37. Circumference 50 ft. 40. Circumference 314.16 ft.

41. How many square yards are there in a floor 24 ft. long and 15 ft. wide?

42. The base of a triangle is 20 ft. and the altitude 18 ft. What is the area?

43. The circumference of a circle is 31.416 ft. and its radius is 5 ft. What is its area?

44. When the diameter of a circle is 50 ft., what is the circumference?

45. When the radius is 6 ft., what is the circumference?

46. When the circumference is 78.54 ft., what is the radius?

47. Mr. Clark's farm is 35 ch. long and 25 ch. wide. How many acres does it contain?

48. A certain field is 70 rd. long and 65 rd. wide. How many acres are there in the field?

49. A piece of land is 65 rods wide. How long must it be to contain 56⅞ acres?

50. How many sods 10 inches square will be required to turf a lawn 100 ft. long and 50 ft. 6 in. wide?

51. A building lot measures 60 feet in front. What must be its depth to contain ¼ of an acre?

52. How many tiles, each 8 inches square, will be required for the floor of a room 24 ft. by 30 ft.?

53. How many shingles will be required for a roof 45 ft. long, and each of its two sides 20 ft. wide, allowing 8 shingles to the square foot?

54. How many square yards of oil-cloth are needed to cover a floor 18 ft. by 24 ft. 6 in. ?

55. A owns a city lot 168 ft. long and 42 ft. wide. He uses ¾ of it for a lawn. How many square yards does the lawn contain ?

56. How long will it take a man to mow the above lot, if it takes him a minute to run a 2-foot lawn-mower lengthwise of the lot ?

57. A pony can reach 40 feet in any direction from the stake to which he is picketed. Over how many square rods of surface can he graze ?

58. What is the diameter of a tree that is 10 feet in circumference ?

59. A basin measures 9 inches across the top and 6 inches across the bottom. How much farther around the top than around the bottom ?

60. How many acres of land are enclosed by a circular mile track ?

61. A landscape gardener lays off a circular grass-plot whose radius is one rod, and near it a semicircular plot having a radius two rods in length. Compare their areas.

62. If the area of a triangle is 9 acres and the base is 80 rods, what is the altitude ?

63. Find the area of one gable end of a building 40 ft. wide, the ridge being 15 ft. above the eaves.

64. Find area of a triangle whose altitude is 14 ft. and its base 12 ft.

65. Find the area of triangle whose altitude is 4 in. and base 12 in.

CARPETING ROOMS.

175. In making a carpet, the carpeting is cut from a roll into strips which are usually laid from end to end of the floor, or lengthwise. Sometimes the strips are laid across the room.

1. How much carpeting must I purchase to cover a room 6 yd. long and $4\frac{3}{4}$ yd. wide, strips running lengthwise?

SOLUTION. — It will be necessary to purchase as much carpeting as if the room were 5 yd. wide, the excess of $\frac{1}{4}$ yd. being turned under in the last strip.
1 strip contains 6 yd. 5 strips = 5 times 6 yd. = 30 yd. *Ans.*

2. How many yards must I purchase, if the strips are laid across the room?

SOLUTION. — 1 strip contains $4\frac{3}{4}$ yd. 6 strips = 6 times $4\frac{3}{4}$ yd. = $28\frac{1}{2}$ yd. *Ans.*

Carpeting is commonly 1 yd. or $\frac{3}{4}$ yd. wide.

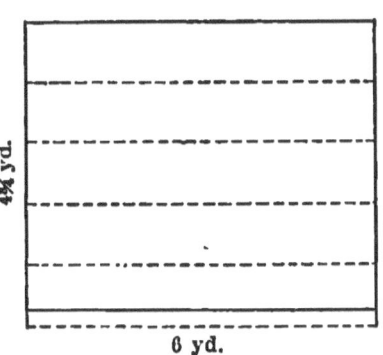

6 yd.

NOTE. — It is often necessary to purchase more than enough carpeting to cover a room, on account of the waste in matching patterns.

This diagram represents the floor in Ex. 1. in which the strips are laid lengthwise. Pupils should draw a similar diagram for each floor.

NOTE. — Carpeting is sold by lineal yards or meters, not by square measure.

3. A merchant bought a roll of carpet containing 74 yd. at 85¢ a yard, and sold it at $1.15 a yard. What was his profit?

4. A roll of carpet ¾ yd. wide contains 60 yards; how many square yards of surface will it cover?

5. How many strips of carpet 3 ft. wide will cover a floor 15 ft. wide? 17 ft wide? 18 ft. wide?

6. How many strips 27 in. wide are required for a floor 12 ft. wide? 14 ft. wide? 16 ft. wide?

7. If, in Example 5, the room is 16 ft. long, how many linear yards of carpet will be needed to cover the floor?

8. If, in Example 6, the room is 19 ft. long, how many yards will be required to cover the floor?

9. How many yards of ingrain carpet ¾ yd. wide will be required for a floor 17 ft. wide and 20 ft. long, strips running across the room?

10. How much will a carpet cost at $.90 a yard to cover a floor 22 ft. long and 15 ft. wide, if the strips run crosswise, and no allowance is made for matching?

11. How many yards of carpet ¾ yd. wide will be required for a floor 20 ft. long and 15 ft. wide, if the strips run across the room?

12. How many yards of carpet 1 yd. wide will be required for a floor 18 ft. long and 14 ft. wide, strips running across the room?

13. How many yards of carpeting are needed to cover a floor 24 ft. long and 17 ft. wide, strips running lengthwise and ¾ yd. wide?

14. If my room is 16½ ft. long and 12 ft. wide, how many yards of carpeting 24 inches wide must I buy, if in cutting 6 inches is allowed on each strip for matching?

15. I wish to have a carpet woven. My room is 21 ft. long and 17 ft. wide; how much carpeting, 34 inches wide, must I order to exactly cover the room, no allowance being made for matching?

MEASUREMENTS.

16. How many yards of carpet 2½ ft. wide will cover a floor 7½ yards long and 14 ft. wide, if strips run lengthwise, and it requires ½ yd. for matching?

17. What will it cost to carpet a room 15 ft. by 17½ ft., with carpet 30 inches wide, at $1.20 per lineal yard, if the strips run lengthwise, and an allowance of 9 in. to each strip be made for matching?

18. How much less would be the cost with no loss for matching?

19. A room 31 ft. by 17 ft. is to be covered with carpeting 30 in. wide. How many yards must be purchased, and how wide a strip must be turned under?

20. At $2.50 a yard, what will be the cost of a carpet to cover a parlor floor 6 yd. long and 5¼ yd. wide, if ¾ yd. is wasted in matching?

21. How many yards of matting 1½ yd. wide will be required for an assembly room 85 ft. 8 in. long and 64 ft. 6 in. wide, strips running across the room?

22. A room 17 ft. 6 in. long, 14 ft. wide, is to be carpeted with carpet ¾ yd. wide. A border ¾ yd. wide goes around the outside. How many yards of border, and how many yards of carpet if strips run lengthwise, and there is a waste of one foot on each strip for matching?

PLASTERING AND PAINTING, ETC.

176. Plastering and Painting are usually done by the square yard. Allowance is sometimes made for doors and windows, which are called openings. Allowance is also sometimes made for base-boards and wainscoting.

1. A room is 18 ft. long, 12 ft. wide, and 10 ft. high. How many square yards in the walls and ceiling, making no allowance for openings?

NOTE. — Let the pupils draw a diagram for each room, representing the four walls in a line. The entire length of the walls will be 2 × (18 ft. + 12 ft.) = 60 ft. The area of the four walls, 60 ft. by 10 ft. = 600 sq. ft.

Area of ceiling 18 ft by 12 ft. = 216 sq. ft. 600 sq. ft. + 216 sq. ft. = 816 sq. ft. = 77⅔ sq. yd., area of walls and ceiling.

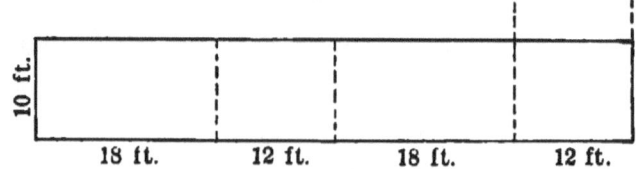

2. Find the cost of plastering the walls and ceiling of a room 35 ft. long, 26 ft. 6 in. wide, and 15 feet high, at $.45 a sq. yd., allowing 1024 sq. ft. for doors, windows, and base-board.

3. How many square yards of plaster in the sides and ceiling of a room 30 ft. long, 24 ft. wide, and 10 ft. high, allowing for a base-board 1 ft. high, 2 doors 3 ft. by 8 ft., and 4 windows 3 ft. by 6 ft. ?

4. Find the cost of plastering the ceiling of a room 18 ft. by 20 ft., at 10 cents a square yard.

5. A room 15 ft. by 18 ft., and 10 ft. high, has 4 doors each 3 ft. by 7 ft., and three windows each 3 ft. by 6 ft. Find the cost of plastering the walls and ceiling of the room at 30 cents a square yard, deducting one-half the surface for openings.

6. How many sq. yds. of plastering in the ceiling of a room 20 ft. long, 9 ft. high, and 15 ft. wide, no allowance for openings ?

7. At $.30 a sq. yd., how much will it cost to plaster a room 21 ft. 6 in. long, 16 ft. wide, and 9 ft. high, the base-board being 8 inches wide, and allowing for 3 windows

7 ft. by 2⅓ ft., and six doors of the same dimensions as the windows?

8. Find the cost of plastering the walls and ceiling of a room which is 36 ft. long, 27 ft. wide, and 9 ft. high, at 25¢ per square yard.

9. My study is 18 ft. long, 16 ft. wide, 8½ ft. high, and contains 1 door 3 ft. by 7 ft., and 2 windows, each 3 ft. by 6 ft. The base-board is 9 in. high. What will it cost, at 36¢ per sq. yd., to plaster it, making full deduction for openings?

10. At 35 cents a sq. yard, what will be the cost of plastering the walls and ceiling of a room 6 yd. long, 5 yd. wide, and 3 yd. high, an allowance of 20 sq. yards being made for openings, etc.?

11. Find the cost of plastering a room 18 ft. square and 10 ft. high, at 25 cents a sq. yard, ⅛ being deducted for openings?

12. A close fence 6½ ft. high surrounds a vacant lot 450 ft. by 380 ft. At 7 cents a sq. yard, what will be the cost of painting both sides of the fence?

13. Find the cost at 18¢ per sq. yard to plaster the sides and bottom of a cistern 8 ft. 6 in. square, and 9 ft. deep.

14. Find the square yards of plastering on a room 20 ft. long, 17 ft. 6 in. wide, 9 ft. high. Allow for 6 windows, each 7 ft. 6 in. high, 3 ft. wide, and 4 doors, each 7 ft. high and 3 ft. 9 in. wide.

PAPERING WALLS.

177. Wall-paper is sold by the roll. A Single Roll is 8 yd. long; a Double Roll, 16 yd. long. Borders are sold by the lineal yard.

The number of rolls needed for a room is found by

dividing the area of the space to be papered by the area of one roll. The width of wall-paper is commonly 18 in.

NOTES. — Unless otherwise stated, a roll is considered as 8 yd. long and 18 in. wide.

Dealers in wall-paper do not sell a part of a roll. If a part of a roll is needed, a whole roll must be purchased.

1. What would it cost to paper the walls of a room 18 ft. long, 12 ft. wide, and 9 ft. high, with paper 8 yd. to the roll and $\frac{1}{2}$ yd. wide, at 45¢ a roll?

2. How many strips of paper, and how many double rolls, will paper the sides of a room 15 ft. long, 12 ft. wide, and 8 ft. high, each roll being $1\frac{1}{2}$ ft. wide, and 16 yd. long, no allowance being made for matching?

3. How many double rolls of paper 16 yd. to the roll, $\frac{1}{2}$ yd. wide, will be required to paper the walls and ceiling of a room 25 ft. long, 10 ft. wide, and 10 ft. high, 110 sq. ft. being deducted for doors, windows, etc?

4. How many rolls of paper will be required for the walls of a room 16 ft. by 20 ft., and 9 ft. high above the base-board, allowing for 3 doors, each 3 ft. by 7 ft., and 3 windows, each 3 ft. by 6 ft.?

5. What will be the cost of the paper and border for the above room at 30 cents a roll for the paper, and 15 cents a yard for the border?

6. How many rolls of paper must be purchased to paper the walls and ceiling of a library, 12 ft. long, 10 ft. 6 in. wide, and 8 ft. high, the base-board being 6 in. wide, and the border $1\frac{1}{2}$ ft. wide, with paper $\frac{1}{2}$ yd. wide and 8 yd. long, the paper extending from the border to the base-board?

BOARD MEASURE.

178. A **Board Foot** is a square foot of the surface of a board, 1 inch thick, or less.

MEASUREMENTS. 83

To find the number of board feet in lumber that is more than one inch thick, we must multiply the number of board feet in the surface by the number of inches in the thickness.

A board 10 ft. long, 1 ft. wide, and 1 in. thick or less contains 10 board feet; but a beam 10 ft. long, 1 ft. wide, and 8 in. thick contains 8 times 10 board feet = 80 board feet.

To find the number of board feet in a tapering board, the average width must be found by taking $\frac{1}{2}$ the sum of the widths of the two ends. Thus a board 10 ft. long, and 12 in. wide at one end, and 6 in. wide at the other, contains as many board feet as if it had a uniform width of 9 inches. $(12 + 6) \div 2 = 9$.

179. The number of board feet = Length (in feet) × Width (in feet) × Thickness (in inches).

NOTE. — When the thickness is one inch or less, the number of board feet is the product of the length and width in feet.

1. How many board feet in a board 15 ft. long, 15 in. wide, and 1 inch thick?

2. How many board feet would there be in the board (Ex. 1) if it were $\frac{3}{4}$ in. thick? 2 in. thick? $1\frac{1}{2}$ in. thick?

3. How many feet of lumber one inch thick will be required for a tight board fence 6 ft. high around a yard 4 rods square?

4. How much lumber (Ex. 3) will be required for an open board fence, 4 boards high, boards 8 inches wide, and 5 inches apart?

5. I need 213 planks 4 ft. 8 in. long, 1 ft. wide, and 2 inches thick, to build a sidewalk. How much will they cost at $13 a thousand?

6. How many feet of lumber will it take to build a line fence 168 **ft. long**, the fence being 5 boards high, and the boards 12 ft. long and 6 in. wide?

7. What will be the cost of 10 planks each 12 ft. long, 10 in. wide, and 3 in. thick at $16 per M.?

8. Find the cost of a stick of timber 8 in. square, and 40 ft. long, at $18 per M.

9. What is the cost of 8 sticks of timber each 36 ft. long, 10 in. wide, 8 in. thick, at $12 per M.?

10. How many board feet of 2-inch planking will it take to make a box 2 ft. 8 in. long, 2 ft. wide, 1 ft. 6 in. deep inside?

11. Find the cost of 7 planks 12 ft. long, 16 in. wide at one end, and 12 in. at the other, at $.08 a board foot.

12. At $18 per M., find the cost of flooring a room 21 ft. by 16 ft., allowing $\frac{1}{8}$ of the lumber for matching.

NOTE.— Find area of floor and add $\frac{1}{8}$.

13. Find the cost of a board 20 ft. long, 22 in. wide at one end, and tapering to 16 in. at the other, and $1\frac{1}{2}$ in. thick, at $30 per M.?

14. At $12 per M., what will be the cost of 2-inch plank for a 3 ft. 6 in. sidewalk on the street side of a rectangular corner lot 56 ft. by 106 ft. 6 in.?

MISCELLANEOUS.

180. 1. My dining-room is 15 ft. long and 12 ft. wide; the walls are 10 ft. high. What will it cost to paper the walls and ceiling with paper $1\frac{1}{2}$ ft. wide, if there are 8 yd. in a roll, and each roll costs $.37$\frac{1}{2}$. ($\frac{1}{10}$ allowed for openings.)

2. What will a carpet for the dining-room (Ex. 1) cost me at $.75 a yd., carpet $\frac{3}{4}$ yd. wide?

3. There are three windows in the dining-room. What will it cost to furnish them with shades at $1.10 each and sash-curtains at 1.37\frac{1}{2}$ each?

4. I bought a table at $14.50, six chairs at $2.75 each, a sideboard for $30, and other furniture for $28.97. I also spent $20 for new table linen; what did it all cost?

5. What was the entire cost of refurnishing my dining-room?

6. What will it cost to carpet a room which is 24 ft. long, and 18 ft. wide, with Brussels carpet 1 yd. wide; no waste in matching, at $1 per yard?

7. Find the cost of plastering sides and ceiling of a room 26 ft. long, $13\frac{1}{2}$ ft. wide, 13 ft. high, at 9 cents a square yard, allowing 25 sq. yd. for openings.

8. What will it cost to build a cement walk 40 ft. long, and 6 ft. wide, at $1.25 per sq. yd.?

9. A field, containing 8 acres, is 60 rd. long. How wide is it?

10. How many yards of carpet will cover a floor 18 ft. long, 16 ft. wide? Carpet one yard wide, strips to run lengthwise of room.

How many yards if the strips run crosswise of the room?

11. Find the cost of fencing a rectangular corner lot 68 ft. by 130 ft., the street fence costing 54 cts. a yard, and the line fences 25 cents a yard, but only half of the cost of the latter to be charged to the lot.

12. Find the cost of a carpet $\frac{3}{4}$ of a yd. wide, at $1.50 per lineal yard, for a room 20 ft. long and 18 ft. wide, strips running lengthwise, and allowing a waste of $\frac{1}{4}$ of a yd. on each strip for matching.

13. What is the breadth of a rectangular lot whose area is 75 sq. ch., and the length 9 ch.?

14. What is the circumference of a circle whose radius is 9 ft.?

15. How many square yards in the above circle?

16. How many revolutions does the 5-foot driving-wheel of a locomotive make in going 30 miles?

17. Find the area of a triangle whose base is 5 feet and altitude 3 ft.

18. If the circumference of the earth is 25000 miles, what is the diameter?

19. The circumference of a circle is 18 ft. What is its radius?

20. How many sq. yards in a triangle whose base is 48 ft. and whose altitude is 24 feet?

21. Find the area of the gable-end of a house whose width is 25 feet and whose ridge is 10 feet 6 inches higher than the base of the gable.

22. If the diameter of the earth is 8000 miles, what is the circumference?

23. How many board feet in 10 planks 18 ft. long, 15 in. wide, and 2 in. thick? and what will they cost at $40 per M.?

24. Find the cost of 10 joists, 3 in. by 12 in., 16 ft. long, at $25 per M.

VOLUMES.

181. Anything that has length, breadth, and thickness is called a **Solid** or **Volume**.

182. A **Rectangular Volume** is a solid having six rectangular faces.

183. A **Cube** is a solid having six equal square faces.

184. A **Cubic Inch** is a cube 1 inch long, 1 inch wide, and 1 inch thick.

185. The **Volume** or **Solidity** of a body is the number of cubic units that it contains.

MEASUREMENTS. 87

1. How many cubic inches in a block 4 in. long, 3 in. wide, and 2 in. thick?

The block is made up of two layers, each 1 in. thick. In the top layer there are 4 times 3 cu. in. In the two layers, therefore, there are 2 × (4 × 3 cu. in.) = 24 cu. inches.

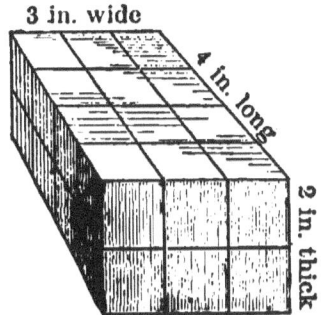

3 cu. in.
4
―――
12 cu. in.
2
―――
24 cu. in.

The multiplier must be considered as abstract. The three dimensions must have the same unit. The length, breadth, and thickness of a rectangular solid are its dimensions.

186. Length × breadth × thickness = Solidity.

Solidity ÷ either dimension = the product of the other two.

Solidity ÷ the product of two dimensions = the other.

2. Find the number of cubic feet of air in a schoolroom 32 ft. square and 12 ft. high.

3. How high is a room that is 24½ ft. long, 20 ft wide, and contains 4410 cu. ft.?

4. A cu. foot of ice weighs 56¼ pounds. How much will a load of 22 cakes weigh, each cake measuring 2 ft. square, and 1 ft. thick?

5. The capacity of a rectangular box is 480 cu. in. The box is 8 in. wide, and 5 in. deep. How long is it?

6. A schoolroom is 25 ft. long, 18 ft. wide, and 12 ft. high. If 60 pupils are seated in it, how many cu. ft. of air are allowed for each child?

7. A man sold 3 blocks of Vermont marble, each 8 ft. long, and 6 in. × 6 in. at the ends. How much did he receive for the marble at $3.50 per cu. ft.?

8. A hot-house bed is 3 ft. 9 in. long, and 3 ft. 4 in. wide, inside measure. How deep must it be to contain 25 cu. ft. of earth, and allow 6 in. for the growth of the plants?

9. How many bricks 8 in. by 4 in. and 2 in. thick will be needed for a wall 60 ft. long, 20 ft. high, and 2 ft. thick, making no allowance for mortar?

10. How many rectangular blocks 12 in. by 8 in. by 3 in. can be packed into a wagon-box 10 ft. long, 4 ft. wide, and 2 ft. 6 in. deep?

11. How many cubic yards of earth must be excavated from a cellar 30 ft. 10 in. long, $21\frac{1}{4}$ ft. wide, and 5 ft. 8 in. deep?

12. How many square feet in the surface of a rectangular box 3 ft. 4 in. long, 2 ft. 2 in. wide, and $1\frac{1}{2}$ ft. high?

13. How many cubes $2\frac{1}{2}$ inches on each edge can be sawed from a block of marble 10 ft. $2\frac{1}{2}$ in. long, 6 ft. 5 in. wide, and 3 ft. 4 in. thick?

14. A box is 1.5 in. long, .85 in. wide, and .58 in. deep. What is its capacity in cubic inches?

15. The altitude of a cylinder is 8 ft. and the circumference of the base is 3 ft. What are the cubic contents of the cylinder?

> NOTES. — Contents of a cyclinder = Area of Base × Altitude.
> Area of curved surface of a cylinder = circumference of Base × Altitude.
>
> This may be seen by cutting a piece of paper so that it will exactly cover the curved surface of a small cylinder.

16. What is the area of the curved surface in the cylinder mentioned in Ex. 15?

17. How much tin will be required to make 2 doz. cylindrical shaped cans, with a diameter of 4 in. and altitude

of 7 in., allowing tin for the curved surface and the two circular ends?

18. How many cubic inches of water will the 2 doz. cans (Ex. 17) contain?

19. What will it cost to dig a cellar 36 ft. long, 24 ft. wide, and 6 ft. deep, at 20 cts. a cubic yard?

WOOD MEASURE.

187. A pile of wood 8 feet long, 4 feet wide, and 4 feet high makes a **Cord**.

One of the 8 feet in length of a cord of wood is a **Cord Foot**.

NOTE. — This may be illustrated by placing side by side 8 books of equal size. One of the books represents a cord foot.

How many cords of wood in the following:

1. A pile 18 ft. long, 4 ft. wide, 8 ft. high?
2. A pile 50 ft. long, 8 ft. wide, 6 ft. high?
3. A pile 19 ft. long, 2 ft. wide, $5\frac{1}{2}$ ft. high?
4. A pile 16 ft. long, $4\frac{1}{2}$ ft. wide, 7 ft. high?
5. What is the cost of a pile of wood 10 ft. long, 4 ft. wide, and 8 ft. high, at $\$4\frac{1}{4}$ a cord?
6. How high must a pile of wood be piled to contain 10 cords, if the pile is 50 ft. long?
7. How many cords of wood can be piled in a shed 24 ft. long, 18 ft. wide, and 12 ft. high?
8. How many cords of building-stone in a pile 18 ft. long, $6\frac{1}{2}$ ft. wide, and 3 ft. high?
9. At $3.50 a cord, what will be the cost of a pile of stone 15 ft. long, $4\frac{1}{2}$ ft. wide, and 5 ft. high?
10. How many cubic feet in a cord of 2-foot wood? 3-foot wood? 18-inch wood?

CAPACITY OF BINS.

188. 1. A bushel fills 2150.42 cubic inches of space. How many bushels of wheat can be contained in a bin 5 ft. × 5 ft. × 4 ft. ?

$$5 \times 5 \times 4 \times 1728 \div 2150.42.$$

NOTE. — A bushel fills 1¼ cu. ft. of space nearly.

2. A wine gallon fills 231 cubic inches of space. How many gallons of water can be contained in a rectangular tank 10 ft. by 8 ft. by 4 ft. ?

NOTE. — A cubic foot of space contains $\frac{1728}{231}$ gal. = 7½ gal. nearly.

Find the contents in bushels:

3. Of a bin 6 ft. long, 5 ft. wide, and 4 ft. high.

4. Of a wagon-box 10 ft. long, 42 in. wide, and 22 in. high.

5. Of a box 3 ft. by 2½ ft. by 2½ ft.

6. How high must a bin 8 ft. long and 5 ft. wide be built to contain 120 bushels?

Find the contents in gallons.

7. Of a tank 8 ft. by 6 ft. by 2½ ft.

8. Of a cistern 6 ft. by 5 ft. by 4½ ft.

9. Of a tank 5½ ft. square and 6 ft. deep.

10. How many barrels of water will a cistern contain that is 6 ft. by 6 ft. by 7 ft. ?

11. A circular cistern is 5 ft. in diameter and 6 ft. deep. How many barrels of water will it hold?

NOTE. — Area of base × altitude.

12. How deep must I build a bin that is 6 ft. square, to hold 90 bushels of wheat?

13. How deep must I build a tank that is 5 ft. square to hold 40 barrels?

LONGITUDE AND TIME.

189. A Meridian is an imaginary line running from the north pole to the south pole.

All places on a meridian have the same time.

NOTE. — The meridians of Greenwich and Washington are the meridians that run through Greenwich and Washington.

190. Longitude is distance east or west from some standard meridian, as Greenwich or Washington. When two places are on the same side of the standard meridian, their difference in longitude is found by subtraction. When on opposite sides, their difference in longitude is found by addition.

1. What is the difference in longitude between two cities, one of which is 20° west longitude, the other 30° east longitude?

$20° + 30° = 50°$. *Ans.*

2. What is the difference in longitude between two places, one of which is 40° E., the other 70° E.?

$70° - 40° = 30°$. *Ans.*

NOTE. — No two places can have a difference in longitude exceeding 180°. If, in finding difference in longitude by addition, the sum exceeds 180°, subtract the sum from 360° to find the true difference.

The earth turns upon its axis from west to east once in 24 hours, thus $\frac{1}{24}$ of its entire circumference, 360°, or 15° of longitude, passes under the sun in 1 hour.

Since the earth turns at the rate of 15° every hour, in 1 minute it turns $\frac{1}{60}$ of 15°, or 15′, and in 1 second $\frac{1}{60}$ of 15′, or 15″. Hence,

The earth rotates 15° in 1 hour, 15′ in 1 minute, and 15″ in 1 second.

3. The difference in longitude between two cities is 18°, 30′. What is the difference in time?

$$\begin{array}{r} 15\,/\,18°\quad 30' \\ \hline 1\text{ hr. }14\text{ min.} \end{array}$$

SOLUTION. — Since the earth turns 15° in 1 hr., 15′ in 1 min., and 15″ in 1 sec., the time can be found by dividing the number of degrees, minutes, and seconds by 15.

4. The difference in time between two cities is 54 min. 19 sec. What is their difference in longitude?

$$\begin{array}{r} 54\text{ min. }19\text{ sec.} \\ 15 \\ \hline 13°\ 34'\quad 45'' \end{array}$$

Since the earth turns 15° in 1 hr., 15′ in 1 min., and 15″ in 1 sec., the distance in degrees, minutes, and seconds may be found by multiplying the number of hours, minutes, and seconds by 15.

STANDARD OR RAILROAD TIME.

191. The railroad companies have divided the country into four time belts, extending north and south. All places in each belt take the time of the meridian which passes through or near the middle of the belt. The belts are as follows: Eastern, Central, Mountain, and Pacific.

The standard meridian for the Eastern belt is the 75th, for the Central belt the 90th, for the Mountain belt the 105th, and for the Pacific belt the 120th.

These standard meridians are 15 degrees apart.

Therefore, when it is noon in the Eastern belt it is 11 A.M. in the Central belt, 10 A.M. in the Mountain belt, and 9 A.M. in the Pacific belt.

In going westward into another time belt, the traveller sets his watch back one hour.

In travelling eastward, he sets his watch ahead one hour.

When it is noon on the standard meridian of each belt, it is called noon at all places in the belt.

NOTE. — Time reckoned by this method is not true solar time, but it secures a uniformity of time which is very desirable.

The time in general use is Railroad or Standard Time.

LONGITUDE AND TIME. 93

Oral.

5. When it is 5 P.M. Mountain time, what is the time in the Pacific belt?

6. When it is 11 A.M. Pacific time, what is the Central time?

7. In travelling from San Francisco to New York, how many times do I change my watch? and do I set it ahead or back?

8. When it is 4 A.M. at Augusta, Me., what is the standard time at St. Louis?

9. When it is 1 P.M. Mountain time at Denver, what time is it at Washington, D.C.?

10. What is the Pacific time at San Francisco when it is 5 P.M. at Chicago?

11. The longitude of St. Paul is 93° 4′ 55″ west, of Philadelphia is 75° 10′ west. What is the difference in longitude?

12. The longitude of New York is 74° 3″ west, of Paris is 2° 20′ 12″ east. What is the difference in longitude?

13. New York City is 74° 4″ west from London. When it is noon at London, what is the true time at New York?

14. The longitude of Boston is 71° 4′ west, and Chicago is 87° 36′ west. Chicago is how far due west from Boston, if there are 51.27 miles in one degree at their latitude?

15. A person travelled until his watch was 3 hours too fast. In what direction and how far did he go?

16. What is the difference in standard time between Boston and Chicago?

17. If a person goes from New York to San Francisco, will his watch be too fast or too slow? and how much?

18. The difference in longitude between two places is 17° 54′ 55″. What is the difference in time?

94 SENIOR ARITHMETIC.

19. The longitude of San Francisco is 122° 26′ 15″ west, and that of Cincinnati is 84° 26′ west. When it is 9 A.M. at San Francisco, what is the time at Cincinnati?

20. The longitude of Boston is 71° 3′ 30″ west, and that of Paris is 2° 20′ 12″ east. When it is 30 min. past 2 P.M. at Paris, what is the time at Boston?

21. Chicago is 87° 38′ west. When it is 27 min. 36 sec. past 11 A.M. at Chicago, it is 10 min. past 12 M. at Washington. What is the longitude of Washington?

22. St. Louis is 90° 15′ 15″ west longitude. A gentleman arriving there from Boston, in 71° 3′ 30″ west, finds that his watch, which was set in Boston, is not right. What change must he make?

23. Mr. Jones started from Philadelphia, and travelled until his watch was 1 hour 30 min. slow. How many degrees did he travel? and in what direction?

24. When it is 12 o'clock noon at Chicago, what time is it in a place 60° 30′ 30″ west of Chicago?

25. Two men start from the same place, and travel in the same direction, one going 3 degrees and the other 5 degrees per day. They travel until their difference in time is 4 hours. How many days are they travelling?

REVIEW OF DENOMINATE NUMBERS.

192. 1. Define a simple number; denominate; compound.
2. For what is linear measure used?
3. For what is square measure used?
4. For what is cubic measure used? } Give tables.
5. For what is liquid measure used?
6. For what is dry measure used?

REVIEW OF DENOMINATE NUMBERS. 95

7. For what is Troy weight used? table. Avoirdupois weight? table. Apothecaries' weight? table.

8. How many grains in a pound Troy? Avoirdupois?

9. How many grains in an ounce Troy? Avoirdupois?

10. What is a long ton? and how used?

11. How many days in a common year? a leap year? What is the solar year? Explain leap year. When does the civil day begin and end?

12. What is the use of circular measure? Define circle, circumference, diameter, radius, arc. Give table. What is the measure of an angle? What is a degree? A quadrant? How do we find circumference? How do we find diameter? What is a right angle?

13. What is a surface? a square? a rectangle? a triangle?

14. Give rule to find area of a square; of a rectangle; of a triangle; of a circle.

15. Define solid, rectangular solid, cube, cylinder. How do we find the volume of rectangular solid? of a cylinder?

16. What is reduction? Reduction ascending? Reduction descending?

17. Define a denominate fraction. Give the different kinds of reduction of denominate fractions.

18. How do we add compound numbers? subtract? multiply? divide?

19. Give the common method of finding the difference between dates. How do we find the exact difference?

20. What is longitude? How do we find difference in longitude between two places on the same side of a prime meridian? On opposite sides?

21. How do we find difference in longitude when difference of time is given? How find difference of time when difference in longitude is given?

22. What is standard time? What are the names of the four time belts? In passing west into a time belt, how does the traveller set his watch? in travelling east?

23. How do we find length when area and breadth are given?

24. How do we find length when volume, thickness, and width are given?

25. What are the dimensions of a rectangular solid?

26. Define cancellation, even number, odd number, prime number, composite.

27. When are numbers prime to each other?

28. How many cubic feet in a cord of wood or stone? How long, wide, and high is a cord of wood? What is a cord foot?

29. For what is board measure used? What is a board foot? Give rule for finding board feet.

30. How do we find the capacity of bins? cisterns?

THE METRIC SYSTEM.

LINEAR MEASURE.

193. The standard unit of Linear Measure in the Metric System is the **Meter**. It is determined by taking one ten-millionth part of the distance from the earth's equator to either of its poles, measured on a meridian. It is equal to 39.37 inches.

QUESTIONS.

194. 1. What denomination in the English linear measure is most nearly like the meter?

2. Draw a line one meter long.

THE METRIC SYSTEM. 97

3. Hold your hands one meter apart.
4. A meter is about how many feet long?
5. How many meters long is your schoolroom? Wide? High?
6. About how many meters in a rod?

HOW THE TABLE IS MADE.

195. Divide a meter into ten equal parts. One of these parts is a **Decimeter**. *Dec* is a Latin stem meaning *tenth*. About how many inches long is a decimeter? Show with your hands the length of a decimeter. What part of a meter is a decimeter?

196. Divide a decimeter into ten equal parts. One of these parts is a **Centimeter**. *Cent* is a Latin stem meaning *hundredth*. What part of an inch is a centimeter? Show its length. How many centimeters in one meter? What part of a meter is a centimeter?

197. Divide a centimeter into ten equal parts. One of these parts is a **Millimeter**. *Mill* is a Latin stem meaning *thousandth*. What part of a meter is a millimeter? How many millimeters in a meter? What part of an inch is a millimeter?

198. Ten meters make one **Dekameter**. *Deka* is a Greek steam meaning *ten*. How many rods in a dekameter? How many feet? How many dekameters long is your schoolroom?

199. Ten dekameters make one **Hektometer**. *Hekto* is a Greek stem meaning *hundred*. How many meters in one hektometer? How many feet long is a hektometer?

200. Ten hektometers make one **Kilometer**. *Kilo* is a Greek stem meaning *thousand*. How many meters in one kilometer? How many feet? What part of a mile?

201. Ten kilometers make one **Myriameter**. *Myria* is a Greek stem meaning *ten-thousand*. How many meters in one myriameter? How many feet? How many miles?

202. These statements may be combined in the following table:

10 Millimeters (mm.)	= 1 Centimeter (cm.)	= .3937+	in.
10 Centimeters	= 1 Decimeter (dm.)	= 3.937+	in.
10 Decimeters	= 1 Meter (m.)	= 39.37+	in.
10 Meters	= 1 Dekameter (Dm.)	= 32.808+	ft.
10 Dekameters	= 1 Hektometer (Hm.)	= 19.927+	rd.
10 Hektometers	= 1 Kilometer (Km.)	= .621+	mi.
10 Kilometers	= 1 Myriameter (Mm.)	= 6.213+	mi.

203.   REDUCTION.

DESCENDING		ASCENDING
1 Myriameter =	1 Millimeter =	
10 Kilometers =	.01 Centimeter =	
100 Hektometers =	.001 Decimeter =	
1000 Dekameters =	.0001 Meter =	
10000 Meters =	.00001 Dekameter =	
100000 Decimeters =	.000001 Hektometer =	
1000000 Centimeters =	.0000001 Kilometer =	
10000000 Millimeters.	.00000001 Myriameter.	

204. The following series of numbers read from the top downward is reduction ascending; read from the bottom upward is reduction descending. All metric numbers may be reduced in this way.

75689132. mm. =
7568913.2 cm. =
756891.32 dm. =
75689.132 m. =
7568.9132 Dm. =
756.89132 Hm. =
75.689132 Km. =
7.5689132 Mm.

All these numbers might be read thus: 7 5 6 8 9 1 3 2.
                                       Mm. Km. Hm. Dm. m. dm. cm. mm.

THE METRIC SYSTEM.

QUESTIONS.

205. 1. How can a metric number be reduced to higher denominations? To lower?

2. Reduce 12345678 mm. to cm.; to dm.; to m.; to Dm.; to Hm.; to Km.; to Mm.

3. Reduce 9.6538714 Mm. to Km.; to Hm.; to Dm.; to m.; to dm.; to cm.; to mm.

4. Reduce 7 Mm. to lower denominations.

5. Reduce 7 mm. to higher denominations.

6. Reduce 6307.1 m. to Km.; to cm.

7. Reduce 31 meters to inches.

8. Write 2 Mm. as meters; 7 Km.; 6 Hm.; 8 Dm.; 5 m. 3 dm.; 2 cm.; 9 mm. Write them all as one number.

9. Reduce 1 Mm. to feet.

10. Write 7 Mm. and 6 mm. in one number, as meters. Reduce it to higher denominations; to lower.

11. Reduce .075 Km. to cm.

12. Reduce 8 Dm. and 6 m. to Mm.; to mm.

13. Write 75 Km. and 62 dm. in one number as meters; as cm.; as Mm.

14. State the value of each figure in 30769.543 M.

15. A ship sails 100 Mm. in one day. How many miles does it sail?

16. Give the table of Metric Linear Measure.

17. Name the standard unit.

18. How is it determined?

19. What is the scale of the Metric system?

20. Name in order the Latin and Greek stems used in the table.

SURFACE MEASURE.

206. The standard unit of surface measure is the **Are** (pronounced like the English *air*).

The Are is a square whose side is one dekameter. It is therefore a **Square Dekameter.**

QUESTIONS.

207. 1. An are is how many meters long? Wide?

2. How many square meters does the are contain?

3. An are is how many inches long? Feet?

4. The are is about how many rods long?

5. About how many square rods does it contain?

6. About how many ares equal one acre?

7. How many ares does the floor of your schoolroom contain?

8. Name all the surfaces you can think of that contain about one are.

TABLE.

208. The table of surface measure, like that of linear measure, is made by prefixing the Latin and Greek stems to the standard unit, thus:

    10 Centares = 1 Deciare, da.
    10 Deciares = 1 Are,   a.
    10 Ares   = 1 Dekare, Da.
    10 Dekares = 1 Hektare, Ha.

NOTE. — The denominations of the above table are little used, except the are, the hektare, and the centare, which are employed chiefly in measurements of land.

209. Draw a square whose side is one meter. How many square meters does it contain? It is how many

THE METRIC SYSTEM. 101

decimeters on a side? How many square decimeters does it contain? How many square decimeters make one square meter?

210. Draw a square whose side is one decimeter. How many square decimeters does it contain? How many centimeters long and wide is it? How many square centimeters does it contain? How many square centimeters in one square decimeter? In the same way find how many square millimeters in one square decimeter.

How many sq. Meters = 1 sq. Dekameter?
How many sq. Dekameters = 1 sq. Hektometer?
How many sq. Hektometers = 1 sq. Kilometer?

211. The answers to the above questions form the following table of surface measure, which is used for all ordinary surface measurements:

100 sq. Millimeters = 1 sq. Centimeter, sq. cm.
100 sq. Centimeters = 1 sq. Decimeter, sq. dm.
100 sq. Decimeters = 1 sq. Meter, sq. m.
100 sq. Meters = 1 sq. Dekameter, sq. Dm.
100 sq. Dekameters = 1 sq. Hektometer, sq. Hm.
100 sq. Hektometers = 1 sq. Kilometer, sq. Km.

QUESTIONS.

212. 1. Which denomination of this table is like the are?

2. Like the centare?

3. Like the hectare?

4. How far to the right must the decimal point be moved to reduce sq. m. to sq. dm.?

5. How many places to the left must the decimal point be moved to reduce sq. m. to sq. Dm.?

6. To reduce sq. mm. to sq. cm.?

7. To reduce sq. mm. to sq. dm.?

8. Reduce 5555 ca. to Ha.

9. Reduce 3333 Ha to ca.

10. A field 134 M. long and 7 Dm. wide, contains how many sq. m. of land?

11. How many ares?

12. How many Ha.?

13. How many sq. Dm.?

14. How many sq. Hm.?

15. How many sq. cm.?

16. How many sq. cm. in an oblong 643 cm. long and 2.5 m. wide?

17. How many sq. mm.?

18. How many sq. Km.?

19. One hectare equals about how many acres?

VOLUME MEASURE.

213. The unit chiefly used in measuring wood and stone is the **Stere** (pronounced *stair*), which is a cube whose edge is one meter. What denomination in the English volume measure is most nearly like the stere? How many cubic meters does the stere contain? How many decisteres? How many centisteres? How many millisteres?

QUESTIONS.

214. A cube whose edge is one meter long contains how many cubic meters? It is how many dm. long? Wide? High? How many cu. dm. does it contain? How many cu. dm. = 1 cu. m.? A cube whose edge is 1 dm. contains how many cu. dm.? How many cm. long is it? Wide?

THE METRIC SYSTEM. 103

High? How many cu. cm. does it contain? How many cu. cm. = 1 cu. dm. ? A cube whose edge is 1 cm. contains how many cu. cm. ? How many mm. long is it? Wide? High? How many cu. mm. does it contain? How many cu. mm. = 1 cu. cm. ?

215. From the answers to the above questions make the following :

TABLE OF VOLUME MEASURE.

1000 cu. Millimeters = 1 cu. Centimeter, cu. dm.
1000 cu. Centimeters = 1 cu. Decimeter, cu. m.
1000 cu. Decimeters = 1 cu. Meter, cu. cm.

QUESTIONS.

216. 1. How may cubic millimeters be reduced to cubic centimeters? To cubic dm. ? To cu. m. ?

2. How many places to the right must the decimal point be moved to reduce cu. meters to cu. millimeters?

3. Reduce 7 cu. meters to cu. millimeters.

4. Reduce 5 cu. millimeters to cu. meters.

5. How many steres in one cu. meter?

6. A pile of wood is 30 dm. long, 3 m. wide, and 18 dm. high. How many cu. meters does it contain?

7. How many steres?

8. How many cu. millimeters?

9. How many cu. centimeters of air in an empty box 2 m. by 12 dm. by 75 cm. ?

10. How many cubic dm. ?

11. How many steres of stone in a wall 30 m. long, 5 dm. thick, and 250 cm. high?

CAPACITY MEASURE.

217. The metric capacity measure takes the place of both the liquid and the dry measure of the English system.

The standard unit of capacity measure is the **Liter** (pronounced *leeter*), which is a cube whose edge is one decimeter.

QUESTIONS.

218. 1. The liter is what part of a meter wide? High? Long?

2. What part of a cubic meter does it contain?

3. About how many inches wide is it? High? Long? About how many cu. in. does it contain?

4. Show with your hands how wide, high, and long a liter is.

5. What denomination of English dry measure corresponds most nearly to the liter?

6. Make a full-sized picture of a liter.

7. What object the size of a liter do you know?

TABLE.

219. The table of capacity is formed similarly to the other metric tables, and is as follows: —

10 Milliliters (ml.)	= 1 Centiliter,	c.
10 Centiliters	= 1 Deciliter,	dl.
10 Deciliters	= 1 Liter,	l.
10 Liters	= 1 Dekaliter,	Dl.
10 Dekaliters	= 1 Hectoliter,	Hl.
10 Hektoliters	= 1 Kiloliter,	Kl.
10 Kiloliters	= 1 Myrialiter,	Ml.

QUESTIONS.

220. 1. How many liters in 1 myrialiter? In 1 ml.?

2. How many milliliters in 1 Ml.?

THE METRIC SYSTEM.   105

3. Reduce 12345678 ml. to higher denominations.

4. Read the above number, giving each figure the name of the denomination it represents.

5. Reduce 154.67 cl. to Kl.

6. Reduce .012346 Ml. to dl.

7. How many liters equal one cubic meter?

8. A bin is 2.5 m. wide, 6.4 m. long, and 17 dm. deep. How many liters of oats will it hold? How many Hl.? How many Kl.?

9. A tank is 3 m. long, and 3 m. wide. How many dm. deep must it be to hold 50 Hl. of water?

10. A stone containing 1 stere, if dropped in a pond, would displace how many liters of water?

MEASURES OF WEIGHT.

221. The **Gram** is the unit of weight. It is equal to the weight of a cubic centimeter of distilled water at its greatest density.

TABLE.

10 Milligrams (mg.)	= 1 Centigram,	cg.
10 Centigrams	= 1 Decigram,	dg.
10 Decigrams	= 1 Gram,	g.
10 Grams	= 1 Dekagram,	Dg.
10 Dekagrams	= 1 Hektogram,	Hg.
10 Hektograms	= 1 Kilogram,	Kg.
10 Kilograms	= 1 Myriagram,	Mg
10 Myriagrams	= 1 Quintal,	Q.
10 Quintals	= 1 Tonneau,	T.
	or Metric Ton.	

QUESTIONS.

222. 1. How many grams in 1 Metric Ton?

2. How many mg. in 1 metric ton?

3. Reduce 1 mg. to T.
4. Reduce 1 T. to mg.
5. Reduce 9876543215 mg. to higher denominations.
6. Read the above number, giving each figure the name of the denomination it represents.
7. Recite the table of weight.
8. Spell the name of each denomination.
9. Reduce 7.42 quintals to centigrams.
10. Reduce 543 mg. to Mg.
11. One gram equals 15.432 grains. How many grains in 1 Kg. ?
12. One pound, Avoirdupois, contains 7000 gr. How many pounds are equivalent to one Kg. ?
13. Mr. Smith weighs 100 Kg.; how many pounds does he weigh ?
14. How many grams does a cubic meter of distilled water weigh ?
15. Would a cubic meter of any other substance weigh the same ? State your reason.
16. How many kilograms of water will a tank 4 m. × 3 m. × 12 dm. hold ?

REVIEW QUESTIONS.

223. 1. How many tables in the Metric System ?
2. Name the standard units in the order in which they have been given. Repeat them until you can say them as rapidly as you can talk.
3. Name the prefixes in the same way.
4. Name and describe the unit of capacity measure; of weight; of length; of volume; of surface.
5. Repeat the tables.

GENERAL REVIEW. 107

6. The stere is the unit of what measure? The meter? The are? The gram? The liter?

7. How can metric numbers be reduced to higher denominations? to lower?

8. How many things are to be committed to memory in the Metric System?

9. What is 39.37? 15.432? 10? These are the only numbers that need be remembered.

GENERAL REVIEW.

224. 1. Define fraction, numerator, fractional unit, terms, reduction of fractions.

2. Change 217 to 20ths.

3. Give the principle upon which reduction of fractions is based. Illustrate.

4. Add $25\frac{1}{2}$, $14\frac{2}{3}$, $7\frac{5}{8}$.

5. Give the rule for reducing fractions to their least common denominator.

6. A man owned $\frac{3}{4}$ of a foundry and sold $\frac{1}{8}$ of his share for $1200; what was the foundry worth?

7. Reduce to simple form $(15\frac{2}{3} - 3\frac{7}{8}) \times (2\frac{1}{3} + 5\frac{3}{4})$.

8. $\dfrac{10\frac{1}{2} \div 4\frac{3}{4}}{6 + 8\frac{7}{8}} = ?$

9. Reduce to least common denominator six thirty-fifths, nine twentieths, and five sixteenths, and arrange the results according to value.

10. A man having $130 used $\frac{5}{8}$ of it; how much of it remained?

11. C and D can do a piece of work in 24 days, D can do it alone in 45 days; how many days will C require to do it?

12. The numerator of a fraction is 6510, the denominator 66495; reduce the fraction to its lowest terms.

13. If 7 be added to each term of the fraction $\frac{2}{8}$, will its value be increased or diminished, and how much?

14. Two men are 140 miles apart, and travel towards each other, one at the rate of $3\frac{1}{8}$ miles an hour, and the other at the rate of $4\frac{1}{3}$ miles an hour; in how many hours will they meet?

15. Define decimal fraction; an account; currency.

16. What will 6827 feet of lumber cost at $10.50 per M.?

17. $8.7625 + 31.735 - 17.382569 = ?$

18. Write in words 365. 8752.

19. Find the cost of 7896 pounds of hay at $16 a ton.

20. Express in figures two hundred sixty-five and five thousand one hundred ten millionths.

21. Change .875 to a common fraction in its lowest terms.

22. How is a bill receipted?

23. Give a rule for dividing a decimal by 10, 100, 1000, etc.

24. Reduce 3.25, 12.364, and .56087 to a common denominator.

25. When will a fraction reduce to a perfect decimal?

26. $7.6875 \div 187.5 \times (5\frac{1}{8} + 2\frac{5}{8}) = $ what?

27. How is the place for the decimal point in the product determined?

28. Give the abbreviations of Creditor and Merchandise.

29. Name the seventh decimal order.

30. James Harris of Syracuse, N.Y., sold for cash to Preston White, on Nov. 4, 1887, 42 lb. of sugar at 10

cents; 3 lb. Y. H. tea at $.60; 4 gal. molasses at $.75; 48 yd. sheeting at $.14; 1 box starch 46 cents; and 8 doz. eggs at $.24.

Make the bill in due form.

31. Define a square; a circle.

32. Write the table of cubic measure.

33. For what purposes are the following used: Troy weight, dry measure?

34. The last war with England commenced June 18, 1812, and ended Feb. 17, 1815. How long did it continue?

35. A jeweller made 3 lb. 2 pwt. 2 gr. of gold into rings weighing 5 pwt. 10 gr. each. How many rings were there?

36. Reduce 2 mi. 6 ch. 3 rd. to links.

37. Reduce to integers of lower denominations £$\frac{2}{3}$, and .25256 T.

38. Change 4 ℥ 5 ʒ 2 ℈ 8 gr. to a decimal of a pound.

39. Find the result of 4.8 bu. + 2$\frac{3}{4}$ bu. + .8125 pk. + 2$\frac{3}{8}$ pk. + $\frac{3}{4}$ bu.

40. A grocer bought 35 casks of molasses, each containing 44 gal. 2 qt. 1 pt. How much did they all contain?

41. A ship in 8° north latitude sailed due south until it reached 12° south latitude; find the distance it sailed in statute miles.

42. Find the value of $\frac{7}{15}$ of a ton.

43. Reduce $\frac{3}{50}$ of a year to integers of lower denominations.

44. Reduce $\frac{2}{3}$ of a lb. Troy to integers of lower denominations.

45. Express 120 rd. 2 yd. 1 ft. 6 in. as the fraction of a mile.

46. Reduce 45 sq. rd. 2 sq. ft. 9 sq. in. to the fraction of an acre.

47. What part of a day are 6 hr. 13 min. 20 sec. ?

48. What part of 4 gal. 2 qt. 1 pt. are 1 gal. 1 qt. 1 pt. ?

49. At 25 cents an ounce, what is the value of 18 oz. 10 pwt. 12 gr. of silver ?

50. How much will it cost to fill a bin with corn at $.45 a bushel, if the bin is 10 ft. square on the bottom and 4 ft. deep ?

51. A cistern measures inside the walls 8 by 6 by 9 ft., and lacks $1\frac{1}{2}$ ft. of being full. How many gallons does it hold ?

52. How many bushels will a box of the same dimensions hold ?

53. How many cords of wood in a pile of wood that is twice the length, height, and width of an established cord ?

54. Find the total weight of 5 car-loads of coal, weighing respectively 14 T. 18 cwt. 63 lb., 17 T. 4 cwt. 85 lb., 13 T. 19 cwt. 26 lb., 15 T. 10 cwt. 43 lb., and 14 T. 7 cwt. 90 lb.

55. How many bricks 8 in. by 4 in. by 2 in. will it take to pave a street $\frac{1}{2}$ mile long and $\frac{1}{80}$ mile wide, laying the brick on the longest narrow face ? How many if they are placed on end ?

56. How much wood in three piles, the first of which contains 10 cd. 6 cd. ft. 4 cu. ft., the second 12 cd. 12 cu. ft., the third 17 cd. 1 cd. ft. ?

57. A family consumes daily 6 lb. 14 oz. of bread. If each loaf weighs 1 lb. 6 oz. and costs 7 cents, how much does bread cost the family for the month of August?

58. Find the sum of $\frac{5}{8}$ mi., $\frac{2}{3}$ fur., $\frac{1}{2}$ rd., and $\frac{2}{3}$ ft.

59. From 6½ mi. take 4 mi. 140 rd. 4 yd.

60. A man has a bin 6 ft. long, 4 ft. wide, 3 ft. deep, ⅔ filled with wheat. If he sells 10 sacks, each containing 2 bu. 1 pk. 5 qt., how much is left?

61. From a piece of land 20 rods long, 180 ft. wide, were sold 4 lots, each 50 ft. wide, 150 ft. long. What part remained?

62. A merchant bought two casks of wine, each containing 41 gal. 3 qt., at $1.80 per gallon. One-seventh of it leaked away. He sold 9 kegs, each containing 5 gal 1 qt. at 30¢ a pint, and the remainder for 40¢ a pint. How much did he gain?

63. A coal-dealer bought 34160 lb. of coal at $2.50 per long ton. He sold 8 loads, each 1 T. 4 cwt. 60 lb., at $3 per short ton, and the rest for $3.25 per short ton. How much did he gain?

64. Change 3,895,504″ to higher denominations.

65. Three quadrants of a circle are equal to how many seconds?

66. Through how many degrees does the minute-hand of a clock pass in 2½ hours? Through how many does the hour-hand pass in the same time?

67. How many minutes elapse between four o'clock Friday afternoon and nine o'clock the following Monday morning?

68. A boy was exactly 10 years old when the United States declared war against Mexico, May 13, 1846. How old was he at the time of the first bloodshed of the Civil War, April 19, 1861?

69. A train leaves New York at six o'clock Monday evening, and travels an average of ⅔ of a mile a minute. When will it reach Buffalo, a distance of 410 miles?

70. A cistern that holds 50 bushels is 6 ft. square; how deep is it?

71. A pile of wood is 6 ft. high and 4 ft. wide. How long must it be to contain 3 cords?

72. A man has a circular garden with a diameter of 36 feet. How many rods of fencing will be required to enclose it?

73. How many square yards in the above garden?

74. A city lot is 35 ft. front and 125 ft. deep. Find the area.

75. A meadow contains $8\frac{3}{4}$ acres. Its width is 35 rods; find the length of it.

76. How many yards of carpeting $\frac{3}{4}$ of a yd. wide will be required for a room 18 ft. wide and 20 ft. long, if the strips run lengthwise, and there is a waste of 6 in. in each strip for matching patterns?

77. The platform in a schoolroom is 30 ft. long and 11 ft. wide. What will be the cost of oil-cloth, at 85 cents a sq. yd., to cover it?

78. How many feet, board measure, in 6 boards 16 ft. long, 10 in. wide, 1 in. thick?

79. Find the cost of 10 Norway sidewalk planks 16 ft. long, 12 in. wide, 2 in. thick, at $18 per M.

80. A class-room is 15 ft. long, 12 ft. wide, 10 ft. high. Find the cost of plastering it at 20 cents a yard.

81. A room is 30 ft. long, 40 ft. wide, and 16 ft. high. Find the number of square yards of plastering in it after making allowance for wainscoting 3 ft. high, 8 windows, 4 ft. by 8 ft., and 6 doors, 3 ft. 6 in. by 7 ft. 6 in.

82. What will it cost to paper a kitchen 12 ft. by 11 ft. and 9 ft. high, with 10-cent paper, if each roll covers 4 sq. yd.?

GENERAL REVIEW. 113

83. Find the cost of papering a room 16 ft. long, 12 ft. wide, 9 ft. 6 in. high, with paper 18 in. wide, 8 yards in a roll, at 50 cents a roll, if 20 sq. yd. be allowed for doors, windows, and base-boards.

84. If a shingle is 4 inches wide, and lays $5\frac{1}{2}$ inches to the weather, how many shingles will it take to shingle one side of a roof that is 32 ft. long by 22 ft. wide, allowing an extra course at the eaves? How many for both sides?

85. What would be the cost of the shingles for both sides of the roof in No. 84 at $3.25 per M.?

86. The product of two numbers is $1\frac{1}{11}$; one of the numbers is $\frac{1}{3}$. What is the other?

87. What fraction multiplied by $\frac{4}{9}$ will equal $\frac{3}{11}$?

88. How many square yards of carpet will be required to carpet a room that is 27 ft. by 33 ft.? How many yards of carpet will be required if the carpet is 30 inches wide?

89. If a hotel uses 3 pounds of coffee a week, what would be paid for coffee at 38 cents a pound for January, February, and March, 1896?

90. When it is noon at New York, 73° 59′ 9″ W., what is the time at Chicago, 87° 36′ 42″ W.? What is the time at New York when it is noon at Chicago?

91. When it is noon at Greenwich, what is the longitude of a place whose time is 8.30 A.M.

92. A and B start at a given point, and travel in opposite directions. A travels until his longitude is 30° 40′ greater than it was, and B travels half as far as A. What is the difference in time between the places they are then in?

93. What part of a pound Avoirdupois is a pound Troy?

94. What part of an ounce Troy is an ounce Avoirdupois?

95. A druggist bought opium at $8 a pound Avoirdupois, and sold it at 75¢ an ounce Troy. What was his profit on 10 pounds?

96. How much heavier is a pound of iron than a pound of gold?

97. What is the difference in the areas of two fields, one being 5 Hm. long and 8 Dm. wide, the other 8 Hm. long and 14 Dm. wide?

98. In a cubic dekameter how many cubic millimeters?

99. A rectangular field is 5.4 Hm. long and 1.5 Hm. wide. How many hektares does it contain?

100. Three fields have an area respectively of 19 A. 146 sq. rd., 12 A. 73 sq. rd. 15 sq. yd., and 9 A. 127 sq. rd. 26 sq. yd. What is the total area?

101. Find the volume and the area of the curved surface of a cylinder whose diameter is 8 in., and whose altitude is 11 in.

PERCENTAGE.

225. Oral.

How much is $\frac{1}{4}$ of 20?

5 is $\frac{1}{4}$ of what?

5 is what part of 20? 5 is how many hundredths of 20?

Questions of Relation may be solved by means of hundredths; thus,

   a. How much is $\frac{25}{100}$ of 20?

   b. 5 is $\frac{25}{100}$ of what?

   c. 5 is what part of 20? 5 is how many hundredths of 20?

Another name for *hundredths* is *per cent;* thus, $\frac{25}{100}$ is 25 per cent, $\frac{8}{100} = 8$ per cent, .16 = 16 per cent, .05 = 5 per cent.

PERCENTAGE. 115

The sign of per cent is %. 25 per cent is 25%, 6 per cent is 6%, ½ per cent is ½%.

Read questions a, b, and c, using the name per cent where necessary.

Write questions a, b, and c, using the sign % in its proper place.

Solve questions a, b, and c, using decimal per cent.

226. Percentage is a process of solving questions of relation by means of hundredths.

Written.

1. How much is 3% of 400?

Solve the above, then form question b, and solve it.

2. How much is 10% of 200?

Solve the above, form question c, and solve it.

Solve the following questions, then form questions b and c, and solve them.

3. How much is 4 per cent of $200?
4. 30% of 500 is how much?
5. How much is 50% of 90?
6. A boy earned $4.00, and spent 10% of it for a book. What was the cost of the book? (Question a.)
7. A farmer had 100 sheep and sold 20% of them. How many sheep did he sell?

Solve the following, then form questions a and c, and solve them.

8. 15 is 10% of what number?
9. 160 is 80% of what number?
10. 50 is 25 per cent of what number?
11. A boy lost 20 cents, which was 5 per cent of all his money. How much money did he have? (Question b.)

12. A farmer sold 150 sheep, which was 50% of his entire flock. How many sheep were in the flock?

Solve the following, then form questions *a* and *b*, and solve them.

13. $20 is what per cent of $100?

14. What per cent of 60 is 15?

15. 40¢ is what % of $4.00?

16. A boy earns $5.00 a week, and saves $2.00 of it. What per cent of his money does he save? (Question *c*)

17. A dealer bought a gross of pencils, and sold 36 of them. What per cent of his pencils did he sell? What per cent remained unsold?

Change the following fractions to others having 100 for a denominator: $\frac{1}{4}$; $\frac{2}{5}$; $\frac{3}{20}$; $\frac{3}{4}$; $\frac{1}{50}$; $\frac{1}{10}$; $\frac{1}{8}$; $\frac{5}{8}$; $\frac{3}{16}$; $\frac{2}{25}$.

Change the above fractions to decimal hundredths.

Read as hundredths: .05; .186; .33$\frac{1}{3}$; .24$\frac{1}{2}$; .27$\frac{1}{4}$; .2725; .5; .1.

Write the above in hundredths as common fractions.

227. Read the following Questions of Relation.

Question *a*. How much is 5% of 200? *Ans.* 10.

Question *b*. 10 is 5% of what? *Ans.* 200.

Question *c*. 10 is what % of 200? *Ans.* 5%.

These three kinds of questions form the basis of a great variety of practical computations, which are classed under the general head of Percentage.

228. Every question in percentage involves three elements: the **Rate** per cent, the **Base,** and the **Percentage.**

The **Rate per cent** is the number of hundredths taken. In question *a*, what is the rate per cent?

The **Base** is the number of which the hundredths are taken. In question *a*, what is the base?

The **Percentage** is the result obtained by taking a certain per cent of a number. In question a, what is the percentage?

How much is 8% of $200?

SOLUTION.— 8% of $200 = 200 × .08 = $16. We now have the three elements, as follows:

8% is the rate, $200 is the base, and $16 is the percentage.
Since $200 × .08 = $16, the percentage;
$16 ÷ .08 = $200, the base;
And $16 ÷ $200 = .08, the rate.

229. Therefore, when any two of these elements are given, the other may be found, thus:

Base × Rate = Percentage;
Percentage ÷ Rate = Base;
Percentage ÷ Base = Rate.

230. Tell which elements are given, and which one is required, in question a; in question b; in question c.

231. Find the percentage and form questions b and c, but do not solve them.

18. 6% of 100 is what?
19. How much is 25% of 200?
20. How much is 40% of 250?
21. What is 4% of 50 men?
22. 20% of 80 is what?
23. 15% of $40 = ?
24. What is 3% of 400 gallons?
25. What is 90% of 200 pounds?
26. 60% of 200 miles = ?
27. 10% of 15 inches = ?
28. What is the base in each of the above questions?

232. Care should be taken to express the decimal rate per cent properly, as hundredths. Every fractional part of 1% must be written at the right of the hundredths place.

$1\% = .01.$ $12\tfrac{1}{2}\% = .12\tfrac{1}{2}$ or $.125.$
$9\% = .09.$ $\tfrac{1}{2}\% = .00\tfrac{1}{2}$ or $.005.$
$10\% = .10.$ $10\tfrac{7}{10}\% = .107.$
$90\% = .90.$ $33\tfrac{1}{3}\% = .33\tfrac{1}{3}.$
$100\% = 1.00.$ $8\tfrac{1}{4}\% = .08\tfrac{1}{4}$ or $.0825.$
$900\% = 9.00.$ $\tfrac{1}{4}\% = .00\tfrac{1}{4}$ or $.0025.$
$125\% = 1.25.$ $\tfrac{1}{8}\% = .00\tfrac{1}{8}$ or $.00125.$

233. Express decimally:

1. 7%	6. $6\tfrac{1}{4}\%$	11. 101%	16. $\tfrac{1}{2}\%$
2. 6%	7. $12\tfrac{1}{2}\%$	12. 110%	17. $\tfrac{3}{4}\%$
3. 2%	8. $15\tfrac{3}{4}\%$	13. 250%	18. $\tfrac{2}{3}\%$
4. 12%	9. $37\tfrac{1}{2}\%$	14. 200%	19. $\tfrac{5}{8}\%$
5. 78%	10. $4\tfrac{5}{8}\%$	15. $127\tfrac{1}{2}\%$	20. $1\tfrac{3}{20}\%$

234. It is often convenient to change the rate per cent to the common fraction form; thus:

$$12\tfrac{1}{2}\% = \frac{12\tfrac{1}{2}}{100} = \frac{\tfrac{25}{2}}{100} = \frac{25}{2} \times \frac{1}{\underset{4}{\cancel{100}}} = \tfrac{1}{8}.$$

Change to common fractions in lowest terms:

1. 25%	5. $16\tfrac{2}{3}\%$	9. 150%	13. $\tfrac{3}{8}\%$
2. 50%	6. $33\tfrac{1}{3}\%$	10. 225%	14. $\tfrac{5}{8}\%$
3. 75%	7. $37\tfrac{1}{2}\%$	11. 175%	15. $\tfrac{3}{4}\%$
4. 20%	8. $87\tfrac{1}{2}\%$	12. 236%	16. $\tfrac{1}{2}\%$

What per cent of a number is $\tfrac{1}{2}$ of it? $\tfrac{1}{3}$? $\tfrac{2}{3}$? $\tfrac{1}{4}$? $\tfrac{3}{4}$? $\tfrac{1}{5}$? $\tfrac{3}{5}$? $\tfrac{7}{10}$? $\tfrac{3}{20}$? $\tfrac{1}{8}$? $\tfrac{3}{8}$? $\tfrac{5}{8}$? $\tfrac{1}{6}$? $\tfrac{5}{16}$? $\tfrac{7}{25}$? $\tfrac{5}{9}$? $\tfrac{1}{15}$? $\tfrac{1}{10}$? $\tfrac{1}{50}$? $\tfrac{9}{10}$? $\tfrac{1}{8}$? $\tfrac{1}{3}$?

PERCENTAGE.   119

235. Express in both the decimal and the common fraction form:

1. 25%	5. $6\frac{2}{3}$%	9. 108%	13. $\frac{3}{4}$%
2. 60%	6. $6\frac{1}{4}$%	10. 150%	14. $\frac{2}{3}$%
3. 18%	7. $7\frac{1}{8}$%	11. 125%	15. $\frac{1}{8}$%
4. 1%	8. $66\frac{2}{3}$%	12. $137\frac{1}{2}$%	16. $\frac{7}{10}$%

236. Per cent is commonly used in the decimal form, but many operations may be much shortened by using the common fraction form.

Solve, using first the decimal, then the common fraction form, and note the difference.

17. How much is 25% of $324?
18. Find $12\frac{1}{2}$% of 960 sheep.
19. What is $16\frac{2}{3}$% of 366 men?
20. Find $33\frac{1}{3}$% of 12 oranges.
21. 50% of 4 tons is what?
22. 20% of $300 = what?

Question a, Oral.

23. What is $\frac{3}{100}$ of 800?
24. What is $\frac{7}{100}$ of 900?
25. $16\frac{2}{3}$% of 48 apples?
26. $33\frac{1}{3}$% of 66 sheep?
27. 50% of 144 men?
28. 20% of 15 eggs?
29. .07 of 500?
30. $12\frac{1}{2}$% of $16?

Written.

237. Rate and base given, to find percentage.

Base × Rate = Percentage.

1. What is 40% of $120?
2. How much is $12\frac{1}{2}$% of 1600 lb.?
3. $18\frac{4}{10}$ of 365 is what?
4. From a flock of 60 sheep, 10% were sold. How many were sold? What is the question in this problem?

5. How much is 100% of 50 bushels?

6. A man having 50 bushels of wheat sold 20 per cent of it. How many bushels did he sell?

7. A man had $1500 in the bank and drew out 40% of it. How much remained in the bank?

Note.—100% represents all he had in the bank. 100% − 40% = 60%, the part that remained. The question then becomes, How much is 60% of $1500?

8. A farmer having 320 acres of land sold 15% of it to one man and 25% to another. How many acres did he sell?

9. A wholesale grocer had 480 bbl. of A sugar, and sold $12\frac{1}{2}$% of it. How much remained unsold?

10. How much is .5% of 80?

11. How much is $\frac{1}{8}$% of $4000?

Question b, Oral.

1. 5 is $\frac{25}{100}$ of what?
2. 5 is 25% of what?
3. 40 is 10% of what?
4. 12 is 8% of what?
5. 30 is $12\frac{1}{2}$% of what?
6. 15 is 50% of what?

Written.

238. **Percentage and rate given, to find base.**

Percentage ÷ Rate = Base.

7. $125 is $12\frac{1}{2}$% of what?
8. 150 bu. is $33\frac{1}{3}$% of what?
9. 240 is 120% of what?
10. $1644 is 40% of what?
11. 75 is $3\frac{1}{4}$% of what?
12. 289 is 50% of what?
13. 25% of my property is $5000. What is the value of my property?

PERCENTAGE. 121

14. I sold a horse for $81, which was 90% of what it cost me. What did the horse cost me?

Question c, Oral.
1. What part of 45 is 15?
2. What per cent of 45 is 15?
3. What per cent of 80 is 60?
4. What per cent of 90 is 30?
5. $40 is what per cent of $60?
6. 12 yd. is what per cent of 36 yd.?
7. 14 bu. is what per cent of 56 bu.?
8. $\frac{2}{3}$ is what per cent of $\frac{1}{4}$?

Written.
239. Base and percentage given, to find rate.
Percentage ÷ Base = Rate.

9. What per cent of $240 is $80?
10. 150 is what per cent of 900?
11. What % of a long ton is a short ton?
12. What % of 5 days is 6 hours?
13. 5 cwt. is what % of 3 tons?
14. $28.16 is what % of $7040?
15. What per cent is $\frac{1}{2}$ of $2\frac{1}{2}$? $\frac{1}{8}$ of $\frac{2}{3}$? $\frac{3}{4}$ of $7\frac{1}{2}$?
16. My salary is $1600 and my expenses $1200. What % of my salary are my expenses?

240. The sum of the base and percentage is called the **Amount**.

241. The difference between the base and percentage is called the **Difference**.

1. Find the amount in the following:
How much is 10% of 20? Find the difference.

2. 20 is what per cent of itself?

3. If 20 is increased by 10% of itself, the amount is 22. What per cent of 20 is 22?

SOLUTION.— The base . . . 20 is 100% of 20.
　　　　　　The percentage . 　2 is 　10% of 20.
　　　　Therefore the amount . 22 is 110% of 20. *Ans.*

4. 100% + 10% = ? 　1 + 10% = ?

5. If 20 is diminished by 10% of itself, the difference is 18. What per cent of 20 is 18?

SOLUTION.— The base . . . 20 is 100% of 20.
　　　　　　The percentage . 　2 is 　10% of 20.
　　　　　　The difference . 　18 is 　90% of 20. *Ans.*

6. 100% − 10% = ? 　1 − 10% = ?

7. What number increased by 10% of itself equals 220?

SOLUTION.— Since 220 is 10% more than the required number, 220 is 110% of the required number.

The amount, 220, is now treated as the percentage, and 110% as the rate; and the question becomes, 220 is 110% of what number? (Question *b*.)

　220 ÷ 1.10 = 200. *Ans.*
　Amount ÷ (1 + rate) = Base.

8. What number diminished by 10% of itself equals 180?

SOLUTION.— Since 180 is 10% less than the required number, 180 is 90% of the required number.

The difference, 180, is now treated as the percentage, and 90% as the rate; and the question becomes, 180 is 90% of what number? (Question *b*.)

　180 ÷ .90 = 200. *Ans.*
　Difference ÷ (1 − rate) = Base.

9. What number increased by 25% of itself equals 290?

10. What number diminished by 25% of itself equals 243?

11. After selling 20% of his sheep, a farmer had 400 sheep left. How many had he at first?

12. The population of a certain city has increased 12% in two years. If it now numbers 56000, what was it at the beginning of the two years?

13. A clerk's salary was increased $6\frac{1}{4}$%. If he now receives $850, what was his original salary?

14. By selling goods at $630 I lose $12\frac{1}{2}$%. What did I pay for them?

15. $580 is 10% less than what number?

16. I sold goods at $450, which was 120% of the cost. What was the cost?

17. After withdrawing 45% of my money from the bank, I still have $1300 on deposit. How much had I in the bank at first?

18. A farmer increased his flock of sheep by $12\frac{1}{2}$%, and then had 900. How many had he at first?

19. A man, after spending a month in the Adirondacks, finds that his weight is 210 pounds, which is an increase of 5%. What was his weight before he went to the Adirondacks?

20. A regiment lost $12\frac{1}{2}$% of its men in an engagement, and had 560 left. How many men were there before the engagement?

21. A owes C $33\frac{1}{3}$% more than he owes B. If he owes C $800, how much does he owe B?

22. 1227.83 is $\frac{1}{2}$% less than what number?

23. $\frac{4}{5}$ is 20% less than what number?

24. A city lot cost $3600, which is 55% less than the cost of the house. What was the cost of the house?

25. A farmer raised 1500 bu. of corn, which was $33\frac{1}{3}\%$ less than the number of bushels of wheat raised. How many bushels of wheat had he?

26. In the year 1896 a merchant's profits were $1836.24, which was 25% more than his profits of 1895. What were his profits in 1895?

PROFIT AND LOSS.

242. Oral.

State the question only.

1. How much is a 10% profit on goods that cost $200?

2. I bought goods for $400, and sold them at a loss of 5%. How much did I lose?

3. If I buy goods at $400, and sell them at $600, what per cent profit do I make?

4. If I buy at $400, and sell at $350, what % do I lose?

5. By selling a house for $1600 I gain $33\frac{1}{3}\%$. What was the cost?

6. John sold his skates for 64 cents, and thereby lost 5%. What did he pay for them?

243. All computations in Profit and Loss come under the rules of **Percentage**.

The *cost* corresponds to the *base*, and the *gain* or *loss* is a percentage of the cost.

The *selling price* is the *amount* when there is a profit, and the *difference* when there is a loss.

244. Written.

7. If I buy eggs for 10 cents a doz., and sell them for $12\frac{1}{2}$ cents, what per cent do I gain?

8. A grocer bought tea at 18 cents per pound, and sold it at 30 cents per pound. What was the rate of gain?

PERCENTAGE.

9. Find the profit on a bicycle that cost $75, and was sold at an advance of 30%.

10. Find the selling price of a horse bought at $88.65, and sold at $3\frac{1}{3}$% below cost.

11. Find the rate per cent of loss on a cow bought for $80, and sold for $60.

12. Find the rate per cent profit on a car-load of Cortland wagons sold for $1090, and bought for $1000.

13. Find the cost of a herd of cattle sold at $12\frac{1}{2}$% above cost at a profit of $240.

14. A man bought books for $194, and sold them at a gain of 32%. What was the gain?

15. I sold a house and lot that cost $11225 at a loss of $5\frac{1}{2}$%. What was the loss?

16. Mr. A., by selling his horse at a profit of 14%, made $32.20. What did the horse cost?

17. By selling sugar at one-half cent per pound profit, a grocer makes ten per cent. What does he get per pound for his sugar?

18. An agent gained $.09 by selling twine 25% above cost. What did it cost him?

19. Find the cost of cotton sold at $16\frac{3}{4}$% above cost at a profit of $211.25.

20. By selling flour at a loss of $14\frac{2}{7}$%, a grocer loses $13.45. What was the cost?

21. A farm that cost $2675 was sold for $3745. What was the gain per cent?

22. Hats that cost $43.50 a doz. are sold for $4.50 apiece. What is the rate of gain?

23. By selling boots for $206.40 a merchant gained 20%. What did they cost him?

24. By selling corn for $92.61, a man gained 12½%. What did it cost him?

25. I sell a horse for twenty per cent less than my asking price, and yet make twenty-five per cent profit. I asked $200. What did the horse cost me?

26. My height is 6 feet 1½ inches, my neighbor is 5 feet 10 inches. What per cent am I taller than he is?

27. A farmer sold 160 acres of land for $2944, which was 8% less than it cost. What did it cost an acre?

28. By selling a horse for $160, I lose 20%. What would have been the selling price had I gained 20%?

29. 14⅞% was gained by selling tea at $.45 a pound. What did it cost a pound?

30. Mr. Brown sold a lot for $4300, and by so doing made 11⅛%. What did he gain?

31. If I buy oranges at the rate of 3 for 3 cents, and sell them at the rate of 2 for 5 cents, what per cent profit do I make?

32. A jeweller sold two watches at $24 each. On one he gained 20%, and on the other lost 20%. What did both watches cost him?

COMMISSION.

245. Oral.

1. A certain agent receives for his services 2% of the value of the goods which he sells. How much will he receive for selling $1000 worth of goods?

2. A purchasing agent receives for his services 3% of the value of goods purchased. How much will he receive for purchasing $2000 worth of goods?

3. How much must I pay my agent for selling $3000 worth of potatoes if I pay him 5%?

PERCENTAGE.

4. At 5% how much will a collecting agent receive for collecting $800?

5. How much must I pay my broker for selling $1000 worth of stocks, if I pay him $\frac{1}{2}$% of their value?

246. An **Agent** is a person who transacts business for another.

247. Some agents are known as **Brokers**, or **Commission Merchants**, according to the kind of business transacted.

248. The compensation of an agent is called **Commission**, or **Brokerage**.

The commission of a purchasing agent is usually a certain per cent of the value of his purchases.

The commission of a sales agent, or of a collector, is usually a certain per cent of the amount collected.

249. The merchandise sent to a commission merchant to be sold is called a **Consignment**.

250. The sender is the **Consignor**, and the person to whom the goods are sent is the **Consignee**.

251. The commission is the percentage, and the amount collected or invested is the base.

252. Written.

6. An auctioneer charges 5% commission for selling $864 worth of goods. What is the amount of his commission?

7. Sold 850 barrels of flour at $5.25 a barrel, and charged 2$\frac{1}{2}$% commission. Find my commission.

8. What is an agent's commission for selling 6840 lb. of butter, at 19 cents a pound, commission 1$\frac{1}{2}$%?

9. A dealer sells real estate for a commission of 2%. How much must he sell during the year to secure an income of $75 per month?

10. A broker in New York received ¼ of one per cent commission for negotiating a sale of 500 one thousand dollar bonds. What was his commission?

11. A real estate man made $50 by receiving 2½ per cent instead of his regular commission of 2 per cent. What did his sales amount to?

12. An agent's fee for collecting bills is 3%. If he receives $86.25 as his commission, how much money has he collected?

13. An agent collected $1864 from a sale of some pictures, and received $4.66 as his fee. What was the rate of commission?

14. An agent having sold 1250 velocipedes at $8 apiece, invested his commission of 1¾% in a new stock company. How many shares at $25 each did he take?

15. What per cent does an agent charge who receives $223 for buying $5575 worth of produce?

16. A man sends his agent $6000 to invest in flour, after deducting his commission of 2%. How much money is spent for the flour, and how much for the agent's commission?

17. A man is paid 5% for collecting $235.75. How much must he pay over to his employer?

18. A merchant sent his agent $3150 with which to buy flour after deducting his commission of 5%. At $4 per barrel how many barrels did the agent buy?

19. An agent sold iron for $9872. He received $163.70, which included a freight charge of $52.64. What rate of commission did he receive?

20. Received as net proceeds from a sale of cotton $1025.70, after paying my agent 2½% for selling. What did the sale amount to?

21. An auctioneer sells 15 tables at $1.45 apiece, 22 chairs at $1.12½ apiece, and some pictures for $8.70, on a commission of 5⅛%. What was his commission, and the net proceeds of the sale?

22. My agent in Boston sold a number of bicycles at $85 each. After deducting his commission of 3⅙%, he returned to me $5759.60. How many bicycles did he sell?

23. An agent who sold 150 lots at $233⅓ each, charged $262.50 for his services. What rate of commission did he get?

24. A collector pays over to his principal $23358.39½ after deducting a commission of 4½%. How much was the entire collection?

25. If I send my agent $367.20, with instructions to buy tea at 30 ct. a pound, and he charges 2% for buying, how many pounds of tea should I receive?

26. A real estate agent charges me two per cent for selling my property in Boston. He remits me $5880. What was his commission?

27. A commission merchant in New York charged $36 for insuring my goods, $14 for cartage, and $50 commission at 2½ per cent for selling them. How much money should he remit to me?

28. Sent my agent $2050 to invest in coal at $4 per ton, after deducting his commission of 2½ per cent. How many tons of coal could he buy?

29. A cotton broker received $2531.71 with which to buy cotton at $.12 a lb. He charged 2¼% commission. How many pounds of cotton did he buy, and what was his commission?

30. A real estate agent receives $162,193.50 from a company to invest in land. If he charges 5% commission, how

many acres of land can he buy at $9 an acre? What is his commission?

31. An agent sold 12000 lb. of cotton at 10 cents a pound. He invested the proceeds in lumber at $25 per M. If his commission for selling was 4%, and for buying 2%, how many feet of lumber did he purchase?

32. A grain-dealer received $4820.40 with which to buy wheat at 60¢ a bushel after deducting his commission of 3%. How much wheat did he purchase?

33. How much stock can be bought for $10827, allowing $1\tfrac{1}{8}\%$ brokerage?

INSURANCE.

253. 1. I keep my house insured for $4000. I pay the insurance company 1% annually. How much do I pay annually?

2. If my house (Ex. 1) burns down at the end of three years, how much shall I receive from the company more than I have paid them?

3. If I pay $75 per annum for insuring my house at 1%, for how much is it insured?

4. At 2% what will be the cost of insuring $10000 of merchandise at $\tfrac{3}{4}$ value?

254. Insurance is security against loss.

255. The payment made for insurance is called the **Premium**. Some of the different kinds of insurance are,

256. Fire, Lightning, Tornado, Accident, Life, and **Marine**.

257. The contract between the insurer and the insured is called the **Policy**.

258. The amount insured is the base, and the premium is the percentage.

5. What will it cost to insure a house worth $2500 at $\frac{2}{3}$ of its value, for three years, at $\frac{3}{4}\%$?

6. Insured a country store for $5,000 and goods for $10,000, at 30¢ on $100. $1 is paid for the policy. What does the insurance cost?

7. What will it cost to insure a mill for $5000, the rate being one and one-half per cent for 3 years?

8. How much will I save by insuring my property for $5000 at $\frac{2}{3}$ of one per cent for 3 years, rather than taking an annual policy for $\frac{1}{4}$ of one per cent?

9. It costs me to insure my house $22.50 when the rate is $\frac{3}{4}$ of one per cent. What is the amount of my policy?

10. A stock of goods is insured for one-half the value, the premium being $30, and the rate $\frac{6}{10}$ of one per cent. What is the value of the goods?

11. The semi-annual premium per one thousand dollars on my $6000 life-insurance policy is $26. What does it cost me a year?

12. A person who pays $12 semi-annually for accident insurance is disabled by an accident for 13 weeks, during which time he receives $10 a week. If he has paid three premiums, how much more does he receive than he has paid out?

13. Paid for insuring a house for $\frac{2}{3}$ of its value, $151. The rate being 75¢ on $100, and the policy costing $1, what was the house worth?

14. To insure a house at $\frac{1}{3}$ of 1% cost me $20. What was the house worth?

15. Paid $18 for insuring goods worth $9000. What was the rate?

16. A merchant pays $75 a year insurance on his stock of goods at $1\frac{1}{2}\%$. What is the value of his stock of goods?

17. A block worth $30000 is insured for ¾ of its value at 2%. How much does the owner lose in case of its total destruction by fire?

18. For how much must a cargo of wheat worth $24000 be insured, at $2\tfrac{1}{2}\%$, so that the owner, in case of loss, may recover both the value of the cargo and the premium.

NOTE. — The value of the wheat = $97\tfrac{1}{2}\%$ of the amount insured.

TRADE DISCOUNT.

259. The deduction of a percentage from the price of merchandise is called **Commercial Discount**.

It is used largely by manufacturers and wholesale merchants. The greatest discounts are for large purchases and cash payment.

260. The **List Price** is the price given in the price-list.

261. The **Net Price** is the list price less the discount.

1. If I can purchase books at 25% off for cash, what must I pay for books listed at $80?

SOLUTION. — 100% − 25% = 75%. 75% of $80 = $60. *Ans.*

NOTE. — When two or more discounts are allowed, the first is deducted, the second computed on the remainder, and deducted from it, etc.

2. At what per cent above cost must a merchant mark his goods so that he may allow a discount of 25% from the marked price, and still make a profit of 10%?

SOLUTION. — Selling price = 110% of cost. This selling price is 75% of the marked price. The question is, 110% is 75% of what. $1.10 \div 75 = 1.46\tfrac{2}{3}\%$, therefore the marked price is $46\tfrac{2}{3}\%$ above cost.

3. Find the sum to be paid on a bill of $264 with 10% off for cash.

4. What is the net price of a bill of goods, the list price of which is $56, subject to discount of 25%?

PERCENTAGE.

5. What must be paid on $935, if 15% and 10% off are allowed?

SOLUTION. — Deducting 15% is the same as allowing 85% of the bill. 85% of 935 = 794.75. 90% of 794.75 = 715.28. *Ans.*

6. Which is the better for the buyer, 40%, or 25% and 15% off?

7. Find a single discount on a bill of $300 equal to 20% and 5% off.

8. A discount of $4 was allowed on a bill, which was then paid with a check for $36. What rate per cent was taken off?

9. Consulting my price-list, I find I can buy goods which are marked $450 at a discount of 20% and 5% off for cash. How much will the goods cost me? and how much discount do I receive?

10. Bought furniture amounting to $520 on credit for 6 months, or 5% discount for cash. What ready money will pay the bill?

11. What is the cash value of a bill of books amounting to $40, on the face of which a discount of 20% and 5% is made?

12. The net amount of a bill of goods is $359.10. What is the gross amount, the rate of discount being 10% and 5%?

13. A set of Encyclopædias, whose catalogue price is $100, can be bought at a discount of 2 tens and 5% off for cash. How much less than the catalogue price will they cost?

NOTE. — The expression 2 tens and 5% means 10%, 10%, and 5%.

14. B offers me some hammocks for $450 with a discount of 20%, and 4% off for cash; and A offers me the same goods at a discount of 2 tens and 4% off. Which is the better offer? and how much?

15. A dealer sold goods at 10% below his asking price, but still made a profit of 20%. What per cent above cost had he marked the goods?

16. A merchant marked carpeting that cost him 60 cents a yard so that he could allow a discount of 10% and still make a profit of 20%. At what price did he mark it?

17. A book-dealer sold a stock of books for $1140, at a discount of 10% from the marked price, and finds that he has made a profit of 14%. What did he pay for the books? and what was their marked value?

18. Find the net amount of a bill for $386 subject to the following discounts, 20%, 10%, and 5%.

TAXES

262. A tax is a sum of money levied upon property and persons for public use.

NOTE.— A tax upon persons is called Capitation or Poll Tax. It is levied in some localities upon men of full age, without regard to their property. It is usually but a small amount upon each person. The practice is going out of use.

263. Property is of two kinds, **Real** and **Personal**.

264. Real Property is immovable property, as lands and buildings.

265. Personal Property is property that is movable, as money, securities, household goods, horses, cattle, etc.

266. A tax assessed upon property is a **Property Tax**.

267. **Assessors** are officers chosen to make a list of taxable property, estimate its value, and apportion the tax.

268. A tax is a percentage upon the assessed valuation of property. The tax on $1 is the rate.

1. The valuation of property in a certain town is $1,500,000, and the rate is $1\frac{1}{2}$%. What is the tax?

2. The tax to be raised in a certain village is $37500. The taxable property is $2,500,000. What is the rate? What will be A's tax on $15000 real estate, and $3000 personal?

3. The property of a town is assessed at $1,250,000. The tax to be raised is $15975. There are 650 polls, assessed at $1.50 each. What is B's entire tax, if his property is assessed at $2500, and he pays the poll-tax?

Rule. — *Deduct the amount of poll-tax, if any, from the whole tax. Divide the remainder by the assessed valuation. The quotient will be the rate.*

To find each person's tax, multiply the assessed valuation by the rate, and to the product add the poll-tax, if any.

4. The officers of a certain town find that all the town expenses for the year 1896 will amount to $46000. The tax-roll shows real estate valued at $2,000,000, and personal property at $300000. What is the rate of taxation?

—5. A certain town votes to raise a tax of $14250, besides the collector's commission of 5%. What is the rate of taxation if the property valuation is $1,000,000?

What is the collector's commission, and what is A's tax, on property valued at $4,500?

6. If the assessed valuation of a village is $2,384,564, and there are 750 polls at $1.50 each, what must be the rate of taxation to meet an expense of $29807.05? What is B's entire tax, if his property is valued at $3875, and he pays for 1 poll?

7. What is the valuation of my property, if my tax, 15 mills on a dollar, amounts to $30.

8. What is my entire tax, if I pay a poll-tax of $1.68. and my property is valued at $24750, when the rate of taxation is $16.28 on $1000?

9. The annual tax-rate for the State of New York for the year 1896 was 2.69 mills on the dollar. The amounts to be raised by tax are as follows: $961116 for general expenses, $4,062,903 for free schools, $2,360,103 for the canals, and $4,368,712 for the State care of the insane. What was the assessed valuation of the property of the entire State?

The tax-rate for 1895 was 3.24 mills. What was the entire tax of 1895?

DUTIES.

269. Duties are taxes on imported goods, levied by the government, and collected at custom-houses. A port containing a custom-house is called a **Port of Entry.**

270. An **Ad Valorem Duty** is a certain rate *on the value* of goods at the place from which they were shipped.

271. A **Specific Duty** is a fixed sum charged upon an imported article, without regard to its value. Allowances are made as follows, in collecting specific duties: for Tare, which is weight of box, cask, etc.; for Leakage, which is loss of liquids in barrels or casks; and for Breakage, which is loss of liquids in bottles.

272. The **Gross Weight** is the weight of articles before any allowances are made.

273. The **Net Weight** is the weight after the allowances are made.

1. At 30 per cent ad valorem, what is the duty on goods valued at $725?

2. What is the duty on 10 gross of silver spoons, valued at $4.50 a dozen, at 30% ad valorem?

3. A. Mark's Sons imported from Lyons 1560 yd. of silk invoiced at 87½¢ per yard. What was the duty at 25¢ a yard, and 30% ad valorem?

4. If the average rate of duty under the McKinley law was 49.58 per cent, and under the Wilson law it is 37 per cent, what is the difference in revenue on $1,000,000 worth of dutiable imports?

5. A merchant bought goods in London invoiced at £450. At the custom-house in New York he paid an ad valorem duty of 18%, and a specific duty of $325. What was the entire cost of the goods in United States money?

6. Imported from England 5 cases of cloths and cashmeres, net weight 95 lb.; value as per invoice £375 10s. What is the duty, the rate being 50¢ per pound, and 35% ad valorem?

QUESTIONS.

274. 1. What is the meaning of the term per cent? How is per cent written?

2. Define base, percentage, rate per cent, amount, difference.

3. Tell how to find percentage when base and rate are given. To find base when percentage and rate are given. To find rate when percentage and base are given.

4. Tell how to find base when amount and rate are given. When difference and rate are given.

5. Define Commission, Brokerage, Insurance, Premium, Policy, Taxes, Real Estate, Personal Property.

6. What is trade discount? List price? Net price? Give rule for finding net price.

MISCELLANEOUS REVIEW OF PERCENTAGE.

275. 1. Find 8% of 750. 6¼% of $12.75. ⅜% of 912. $\frac{5}{16}$% of 2140.

2. A man gave his son 42% of his money, his daughter 25% of it, and his wife 16⅔% of the remainder. If the son received $9350 more than the daughter, what did each receive?

3. A dealer sold a horse and carriage for $637, which was 40% more than cost. If the horse cost ¾ as much as the carriage, what did each cost?

4. What per cent is gained when one-half an article is sold for what the whole cost? When ⅝ of an article is sold for what one half-cost?

5. A merchant pays $35 for a suit of clothes. What must he ask for it, so that he may drop 16% from his asking price, and still make 20% on the cost?

6. 14.35 is $\frac{7}{16}$% of what number?

7. A man spent 20% of his salary for board and 15% of what was left for clothes. If he spent $132 more for board than for clothes, how much did he spend for each?

8. What number diminished by 16⅔% is 605?

9. What number increased 35% is 382.5?

10. Sold a load of wheat weighing 3240 lb. at 68¢ a bushel of 60 lb., thereby making a profit of 6¼ per cent. Required the cost of the wheat?

11. On a certain day the sun rose at 5 o'clock and 43 minutes, and set at 6 o'clock and 25 minutes. What per cent of the day was in sunlight?

12. The salary of a certain teacher of arithmetic is $1600. His real estate tax is $90; his water tax is $25; gas bill, $15; coal bill, $45; other expenses, $325. What per cent of his salary does he save in the bank?

13. The Oswego Starch Factory employs 700 operatives. The population of Oswego numbers 22000. What per cent of the population is employed in the starch factory?

PERCENTAGE.

14. $\frac{5}{8}$ is 25% more than what fraction?

15. A dealer lost 10% of his capital, then gained 20% of the remainder, when he had $2160. How much had he at first?

16. Goods bought for $400 are marked to sell at an advance of 40%, but are finally sold at a reduction of 25% from the marked price. What is the per cent of gain? What is the gain?

17. An article is sold for $2.80, this being an advance of 25%. Find the cost.

18. A merchant buys sugar at an average price of 4 cents a pound, and sells at a profit of 8%. How many pounds must he sell to clear $500?

19. If by selling an article for 59 cents a dealer gains 10% more than by selling for 55 cents, what is the original cost?

20. 15% of an estate is invested in city bonds, 40% in real estate, 25% in railroad stock, and the remainder, $5000, is deposited in a bank. What is the estate worth?

21. Define percentage, base, profit and loss, commission.

22. Give the five formulas of percentage.

23. Express as a decimal $\frac{5}{8}$ per cent.

24. A man having a yearly income of $1500 spends 80% of it the first year, 75% of it the second year, 62½% of it the third. How much does he save in 3 years?

25. 25% of 200 bushels is 2½% of how many bushels?

26. A man sold 80 acres of land for $1472, which was 8% less than it cost. What did it cost an acre?

27. What terms in Profit and Loss correspond to base and amount?

28. Find the cost of fruit sold for $207.70, at a gain of 15%.

29. At what price must hats that cost $1.12 each be marked in order to abate 5%, and yet make 25% profit?

30. What is the base in commission?

31. A commission merchant sells 225 bu. of corn at $.65 a bushel, and 360 bbl. of apples at $2.40 per barrel; commission 5%. Find the commission and the net proceeds.

32. The net proceeds are $3800, the rate 10%. Find the amount of sales and the commission.

33. Find the rate, the commission being $125, and the sum invested $2500.

34. A merchant owning ⅔ of a cargo valued at $44000 insures ¾ of his share at 2½%. What premium does he pay?

35. A man having $400 paid 62½% of it for a carriage. How many dollars had he left?

36. An agent charged $432.46 for selling goods at $49424. What was his rate of commission?

37. A man sold four horses for $100 each; on two he gained 25%, and on the other two he lost 25%. Did he gain or lose on the transaction? and how much?

38. If ⅘ the number of girls in a certain school exceed the boys 10%, and the girls number 275, what is the number of boys?

39. A farmer's sheep increased 10% each year for 2 years, when he had 242. How many had he at first?

40. My New York agent buys for me 40 pieces of silk, 32 yards in a piece, at $5 a yard. He charges 1¼% commission. How much money will it require to purchase the silk and pay his commission?

41. A commission merchant in Boston has sold goods for me to the amount of $6932. He has charged 1¼% commission, $18.50 cartage, and $12.15 for storage. How much is due me?

PERCENTAGE.

42. A boy bought oranges at the rate of 3 for 5 cents, and sold them at the rate of 2 for 5¢. What was his rate of gain?

43. Ten per cent of a number is 32 less than eighteen per cent of the same number. What is the number?

44. I paid $28.87½ for insuring my house for $3850 for three years. What was the rate of the yearly premium?

45. A stock of goods valued at $6300 is insured for ¾ its value at ¾%. What will be the owner's loss if the goods are totally destroyed by fire?

46. A man's income is $1720, which is $16\tfrac{2}{3}\%$ of the sum he has invested. What sum has he invested?

47. From a cargo containing wheat, 1620 bu., or 7%, was washed overboard. What number of bushels remained?

48. A stock-dealer sold 38 head of cattle, which was 4% of his entire herd. How many had he left?

49. In an orchard containing 820 trees, 20% of them were pear-trees, and the remainder were plum-trees. How many plum-trees were there in the orchard?

50. From a cask of wine containing 65 gal. all but 15% was sold. How many gallons were sold?

51. In a school containing 875 pupils 32% of them are boys and the remainder girls. How many girls are there?

52. There is a loss of $500 on a house and lot sold for $5000. What is the per cent of loss?

53. An agent reports that he invested the money remitted him in wheat, which he sold at an advance of 15%; then investing the proceeds in a second quantity, he was forced to sell at a loss of 12½%. He now deducts $100 for expenses and commission, and remits $5333.75 to his employer as the balance due him. Find the loss to the employer.

54. $90 are paid as premium for insuring a block for three-fourths of its value. If the rate of insurance is ¾%, what is the value of the property?

55. New York State has a population of 5,998,000, and New York City has 1,515,000. What per cent of the population of the State live in New York City?

56. During the war of 1861–1865, the State of New York paid $40,000,000 in bounties to her volunteers. Her population at that time was (in round numbers) 4,000,000. What was the average cost to each inhabitant?

57. If I sell 6 horses for what 8 horses cost, what is my rate of gain?

58. Sold wheat for $73.54¼, by which a gain of 15% was made. What did the wheat cost? and what sum was gained?

59. In a school containing 1160 pupils, 638 are girls and the remainder are boys. What per cent are boys?

60. A hall is 42 ft. wide, and 294 ft. long. What per cent of the length is the width?

61. In a certain battle 22⅔% more than ⅓ of the soldiers were killed. If the loss was 110 men, what was the original number?

62. In an orange-grove 8⅓ % of the trees were ruined by frost. If 1100 remained uninjured, how many were destroyed?

63. A merchant sold a lot of goods for $550, thereby gaining 10%. Find the cost of the goods.

64. A man sold a watch for $32, thereby losing 20% on the cost. Find the cost.

65. If a man owns 66⅔ per cent of a factory, and sells 33⅓ per cent of his share for $1800, what is the value of the factory?

SIMPLE INTEREST. 143

66. A sold 30% of his steamship to B; B sold 60% of his purchase to C; C sold 75% of his share to D for $27000. What was the value of the vessel? What was each one's share in dollars after the sales had been made?

67. After a discount of 30% had been made upon the catalogue price of a book, it was sold for $1.75. What was the catalogue price?

68. Bought a horse for $120, and sold him for $135. What part of the cost was the gain? What per cent?

69. Bought tea at 60 ct. a pound. What must I ask per lb. so as to abate 10% and still make a profit of 25%?

70. A merchant's profits for 1895 were $3402.84. If they were 6⅜% less than in 1894, what were they in 1894?

71. In one week John solved 75 problems correctly. If he failed in 16⅜% of the number attempted, how many were there in all?

SIMPLE INTEREST.

276. 1. I borrow $500 for 1 year, and at the end of the year I repay the money and 6% for the use of it. How much do I pay for the use of $500?

2. How much must be paid for the use of $50 for 1 year at 5%? at 7%?

3. How much at 5% per annum must I pay for the use of $1000 for 1 year? For 3 years?

4. I loan James Barnes $500 at 6%. At the end of 2 years he pays me in full. How much does he pay me?

Money that is paid for the use of money is called **Interest**. The money for the use of which interest is paid is called the **Principal**, and the sum of the Principal and interest is called the **Amount**.

Interest at 6% means 6% of the principal for 1 year.

12 months of 30 days each are usually regarded as a year in computing interest.

Oral.

5. What is the interest of $100 for 3 years at 6%?

SOLUTION. —
$100 Principal.
.06 Rate.
$6.00 Interest for 1 year.
3
$18.00 Interest for 3 years.

6. What is the interest of $80 at 5% for $2\frac{1}{2}$ years?

7. What is the interest of $1000 at 5% for 2 yr. 6 mo.?

8. What is the interest of $100 at 6% for 1 year? For $1\frac{1}{2}$? For 2 yr. 6 mo? For 3 yr. 3 mo.? For 1 yr. 6 mo.?

When the time does not include days, find interest as follows:

Principal × Rate × Time = Interest.

9. What is the interest of $297.62 for 5 yr. 3 mo. at 6%?

SOLUTION. —
$297.62
.06
$17.8572
$5\frac{1}{4}$
44043
892860
$93.75 Ans.

NOTE. — Final results should not include mills. Mills are disregarded if less than 5, and called another cent if 5 or more.

Find the interest of:

10. $384.62 at 6% for 2 yr.
11. $463.75 at 7% for 3 yr.
12. $250.50 at 8% for 5 yr.
13. $685.20 at 4% for 6 yr.
14. $596.15 at 5% for 2 yr. 3 mo.
15. $386.42 at $5\frac{1}{3}$% for 6 yr. 5 mo.
16. $950.16 at 10% for $4\frac{1}{2}$ yr.
17. $283.25 at 6% for 2 yr. 8 mo.

SIMPLE INTEREST.

Find the amount of:

18. $284.10 for 3 yr. 2. mo. at 7%.
19. $364.24 for 1 yr. 1 mo. at 6%.
20. $282.50 for 5 yr. 9 mo. at 5½%.
21. $298 for 4 yr. 3 mo. at 6%.
22. $389 for 7 yr. 10 mo. at 5%.
23. $894 for 5½ yr. at 5½%.
24. A man buys a house and lot for $2800. He pays ⅜ of the amount in cash, and the remainder after 1 yr. 4 mo. with 5% interest. Find the amount of the second payment.
25. Required the simple interest and amount of $787.875 for 7 yr. 7 mo. at 7%.
26. Find the interest on a note for $12500 for three months at 8%.
27. A man paid his city tax five months after it became due. His tax was $560. In accordance with city ordinance, 1% is added for each ½ month the taxes are overdue. He pays to the city collector of taxes, who adds 5% collection fee. How much did he have to pay?

THE SIX PER CENT METHOD.

277. By the 6% method it is convenient to find first the interest of $1, then multiply it by the principal.

1. If $.09 is the interest of $1 for a certain time, what is the interest of $2 for the same time? of $10? of $25?

2. The interest of $1 at 6% for a certain time is $.034. What is the interest of $36.25 for the same time?

EXPLANATION. — The interest of $36.25 is $36\frac{25}{100}$ times the interest of $1.

At 6% the interest of $1 for 1 year = ... $.06
        for 1 month = $\frac{1}{12}$ of $.06 = $.00½
        for 1 day = $\frac{1}{30}$ of $.00½ = $.000⅙

3. What is the interest of $50.24 at 6% for 2 yr. 8 mo. 18 da.?

SOLUTION. —

$$\begin{aligned}
\text{The interest of \$1 for 2 yr.} &= 2 \times \$.06 &&= \$.12 \\
\text{for 8 mo.} &= 8 \times \$.00\tfrac{1}{2} &&= .04 \\
\text{for 18 da.} &= 18 \times \$.000\tfrac{1}{6} &&= \underline{.003} \\
\text{The interest of \$1 for 2 yr. 8 mo. 18 da.} &&&= \$.163 \\
\text{The interest of \$50.24 is 50.24 times \$.163} &&&= \$8.19
\end{aligned}$$

4. What is the interest of $1 for 2 months? For 6 days?

Rule. — *Find the interest on $1 for the given time, and multiply it by the principal, considered as an abstract number.*

Or, multiply the number of dollars by the number of days, and divide by 6. The quotient will be the interest in mills.

Find the interest at 6% of:

5. $382 for 6 mo. 24 da.
6. $58.63 for 1 yr. 5 mo. 17 da.
7. $256 for 3 yr. 5 mo.
8. $249.83 for 1 yr. 2 mo. 15 da.
9. $51 for 236 da.
10. $74 for 2 mo. 19 da.
11. $1500 for 1 yr. 3 da.
12. $287.15 for 2 yr. 11 mo. 22 da.

Interest at any rate per cent may be found as follows:
At 7%, find interest at 6%, and add $\tfrac{1}{6}$ of itself.
At 5%, find interest at 6%, and subtract $\tfrac{1}{6}$ of itself.
At 8%, find interest at 6%, and add $\tfrac{2}{6}$ or $\tfrac{1}{3}$ of itself.
At 4%, find interest at 6%, and subtract $\tfrac{2}{6}$ or $\tfrac{1}{3}$ of itself.
At 5½%, find interest at 6%, and subtract $\tfrac{1}{4}$ of itself.

SIMPLE INTEREST. 147

Find the interest and amount of the following:
13. $2350 for 1 yr. 3 mo. 6 da. at 5%.
14. $125.75 for 2 mo. 18 da. at 7%.
15. $950.63 for 3 yr. 17 da. at $4\frac{1}{2}$%.
16. $336.48 for 90 da. at $7\frac{1}{2}$%.
17. $738.53 for 2 yr. 2 mo. 24 da. at 8%.
18. $5000 for 6 mo. 19 da. at 4%.
19. $867.35 for 1 yr. 3 mo. 27 da. at 9%.
20. $260.50 for 5 yr. 21 da. at 10%.
21. $3050 for 3 yr. 3 mo. 3 da. at 12%.
22. $625.57 for 1 yr. 2 mo. 15 da. at 3%.

23. A grocer's bill for $84.36 is paid 8 mo. 12 da. after it becomes due, with interest at 5%. How much is paid?

24. Find the interest at 7% on $37200 for 5 days.

25. A note for $125 was dated March 1, 1894. What was due Aug. 5, 1895, int. at 6%?

26. Find the amount of $460.50 for 2 yr. 7 mo. 15 da. at 5%.

27. What is the amount of a note for $360 due in 3 mo., interest at 5%?

278. On short-time notes, it is customary to compute interest for the actual number of days, using the 6% method.

Find the amount of:
28. $684.23 from June 5, 1895, to July 23, 1895, at 6%.
29. $846 from Jan. 6 to March 9, 1896, at 5%.
30. $2064.28 from April 13, 1894, to June 3, 1894, at 8%.
31. $1428 from May 12, 1892, to June 9, 1892, at 6%.
32. $324 from April 1, 1896, to June 4, 1896, at 7%.

33. $3500 from Feb. 9, 1895, to March 12, 1896, at $4\frac{1}{2}\%$.

34. $862.15 from May 25, 1893, to July 22, 1893, at 6%.

35. What is the amount of a note of $384.16 at 6%, given June 11, 1896, and paid Aug. 12, 1896?

36. A note of $395.80 dated April 5, 1896, was paid Aug. 4, 1896. What was the amount?

37. On Dec. 9, 1894, John Smith borrowed $484, agreeing to pay interest at 5%. He paid the debt in full on March 3, 1895. What did he pay?

38. What is the amount of $58.24 at 7% from April 23, 1893, to July 22, 1893?

39. A bill of $312 with interest at 5% was paid at the end of 90 days. What was the amount?

40. What is the interest of $30000 at 8% for 7 days?

279. Find the interest, using the best method.

	PRINCIPAL.	TIME.	RATE.
41.	$364,	3 yr.,	8%.
42.	$692.15,	1 yr. 3 mo.,	9%.
43.	$342,	62 da.,	6%.
44.	$243.50,	2 yr. 5 mo. 18 da.,	7%.
45.	$.392,	1 yr. 3 mo. 15 da.,	4%.
46.	$150.16,	7 yr. 2 mo. 27 da.,	$4\frac{1}{2}\%$.
47.	$284.10,	1 yr. 8 mo. 18 da.,	6%.
8.	$1400,	2 yr. 1 mo. 12 da.,	7%.
49.	$124,	5 yr. 3 mo. 29 da.,	6%.
50.	$48,	33 da.,	6%.
51.	$124,	112 da.,	5%.
52.	$315,	45 da.,	$4\frac{1}{2}\%$.
53.	$214,	93 da.,	8%.

SIMPLE INTEREST. 149

280. Find the amount of:

54. $365 from April 1, 1895, to July 5, 1897, at 6%.

55. $250 from July 3, 1891, to April 21, 1893, at 9%.

56. $582 from Sept. 4, 1896, to July 8, 1897, at 8%.

57. $346.18 from May 10, 1893, to March 10, 1895, at 6%.

58. $287 from Jan. 1, 1895, to July 1, 1897, at $4\frac{1}{2}$%.

59. $1684 from July 17, 1896, to Sept. 5, 1898, at $7\frac{1}{2}$.

60. $2500 from April 16, 1873, to Oct. 11, 1881, at 5%.

61. $186 from Feb. 12, 1896, to March 4, 1896, at 6%.

62. $346 from March 11, 1895, to Feb. 11, 1896, at 6%.

EXACT INTEREST.

281. When the time includes days, interest computed by the 6% method is not strictly exact, by reason of using only 30 days for a month, which makes the year only 360 days. The day is therefore reckoned as $\frac{1}{360}$ of a year, whereas it is $\frac{1}{365}$ of a year.

Rule. — To compute exact interest, find the exact time in days, and consider 1 day's interest as $\frac{1}{365}$ of 1 year's interest.

1. Find the exact interest of $358 for 74 days at 7%.

SOLUTION. — $358 × .07 = $25.06, 1 year's interest. 74 days' interest is $\frac{74}{365}$ of 1 year's interest. $\frac{74}{365}$ of $25.06 = $5.08. *Ans.*

Find the exact interest of:

2. $324 for 15 d. at 9%.

3. $253 for 98 d. at 4%.

4. $624 for 117 d. at 7%.

5. $153.26 for 256 d. at $5\frac{1}{2}$%.

6. $620 from Aug. 15 to Nov. 12 at 6%.

7. $540.25 from June 12 to Sept. 14 at 8%.

8. $7560 for 90 days at 5½%.

9. Find the exact interest at 5% on a note dated Jan. 14, 1896, and paid March 31, 1896, for $832.

10. Find the exact interest on $800 for 219 days at 4½%.

11. A city treasurer deposits $387,913.56 in the banks at 2% per annum. What interest will the city receive in 5 days?

12. On June 4, 1895, a coal-dealer bought of the D. L. & W. R. R. 235 tons of chestnut coal at $4.10 per ton. At 6% what will be the exact interest on the amount on Jan. 1, 1896?

PROBLEMS IN INTEREST.

282. To find the Rate, when Principal, Interest, and Time are given.

1. What is the rate when the interest of $250 for 4 years is $60?

$10 / $60
6 times 1% = 6%

SOLUTION. — The interest on the principal at 1% for 4 years = $10. Since $10 is the interest at 1%, $60 must be the interest at as many times 1% as $10 is contained times in $60, which are 6 times. Therefore the rate is 6 times 1% = 6%.

Rule. — Divide the given interest by the interest of the principal for the given time at 1%.

2. A man borrowed $4625 for 5 yr. 8 mo. 18 da., and paid $1586.37½ for the use of it. What was the rate of interest?

3. If $30.44 is paid for the use of $960 for 7 mo. 18 da., what is the rate per cent?

4. At what rate per cent must $1450 be loaned for 4 yr. 5mo. to yield $576.37½?

SIMPLE INTEREST. 151

5. At what rate will $1730 amount to $2048.32 in 4 yr. 7 mo. 6 da.?

6. 4 yr. 7 mo. 6 da. after its date a note for $1730 amounted to $2048.32. What was the rate of interest?

7. At what rate % must $5600 be invested for 1 yr. 4 mo. to bear $560 interest?

—8. A Kansas farmer has a mortgage on his farm for $1250. What rate of interest does he pay, if the interest for 2yr. 6mo. equals ⅕ of the debt?

9. At what rate must $2800 be invested to yield a semi-annual interest of $112?

10. At what rate will $600 yield $198 in 5 years and 6 months?

11. At what rate will any sum double itself in 20 yr.?

283. To find Time, when Principal, Interest, and Rate are given.

1. In what time will $250 gain $60 at 6%?

SOLUTION. — The interest of $250 for 1 year at 6% = $15. Since $15 is the interest for 1 year, $60 is the interest for as many years as $15 is contained times in $60 = 4 years.

Rule. — Divide the given interest by the interest of the principal for 1 year.

2. In what time will $600 yield $91.50 interest at 6%?

$600 × .06 = $36, interest on principal for 1 year.
$91.50 ÷ 36 = 2.5416+ years. Reducing the decimal part
of the time to months and days, we have 6 mo. 15 da.
The answer is 2 yr. 6 mo. 15 da.

NOTE. — A decimal less than .5 of a day is not counted, but .5 or more is counted another day.

3. In what time will $530 gain $92.75 interest at 5%?

4. In what time will $400 yield $55 interest at 5½%?

5. In what time will $500 gain $15 at 6%?

6. In what time will $4625 yield $1586.38 at 6%?

7. In what time will $1730 amount to $2048.32 at 4%?

8. The face of a note was $960, rate of interest 5%, and the interest $30.44. How long did it run?

9. I borrowed $1284 at 4½%, and kept it until it amounted to $1421.067. How long did I keep it?

10. For how long will $2700 have to be invested to amount to $2976.25 at 5%?

11. A man received $9.73 interest on $556 at 7%. What was the time?

12. In what time will any sum double itself at 6%?

284. To find **Principal**, when **Interest** or **Amount, Rate,** and **Time** are given.

1. What principal at 6% will gain $60 interest in 4 years?

$.24 / $60.00 (250.
  48
  ───
  120
  120
  ───
    0

SOLUTION. — Since 1 dollar in 4 years will gain $.24 interest, it will take as many dollars to gain $60 interest as $.24 is contained times in $60, or $250.

Rule. — *Divide the given interest by the interest of $1 for the given time and rate.*

2. What principal at 6% will amount to $310 in 4 years?

SOLUTION. — Since $1.24 is the amount of $1 for 4 years, $310 must be the amount of as many times $1 as $1.24 is contained times in 310 = $250.

SIMPLE INTEREST.

3. What sum invested at 5% will give a yearly income of $500?

4. What principal will yield $25 in 6 mo. at 5%?

5. What principal in 3 yr. 6 mo. at 5% will yield $92.75 interest?

6. What sum of money will produce $1586.37½ in 5 years, 8 mo. 18 da. at 6%?

7. What principal will yield $318.32 in 4 yr. 7 mo. 6 da. at 4%?

8. What principal will pay $1556.77½ interest in 2 yr. 9 mo. at 4½%?

9. The amount is $1093.92¼, time 2 yr. 3 mo. 27 da., rate 5%. What is the principal?

10. It required $407.65 to pay a loan at 8% for 7 mo. 24 da. What sum was loaned?

PROMISSORY NOTES.

285. A **Promissory Note** is a written promise to pay a sum of money at a certain time.

286. At least two parties must be named in the note, the **Maker** and the **Payee**.

The Maker makes the promise to pay to the Payee the sum named in the note. This sum is called the **Face**. The owner of a note is called the **Holder**.

Each State has a lawful or *legal rate* of interest.

If no rate is fixed in the note, the legal rate is understood.

287. Interest higher than the legal rate is **Usury**.

288. A note is **Negotiable** when payable to the bearer, or to the order of the payee. It is called negotiable because it can be negotiated; i.e., bought and sold.

SENIOR ARITHMETIC.

289. The two forms of notes given below are negotiable.

NOTE 1.

$510 \tfrac{36}{100}.   Rochester, N. Y., Jan 8, 1896.

Two months after date, *I* promise to pay to the order of ⸺ *Charles M. Warner* ⸺ *Five Hundred ten* and $\tfrac{36}{100}$ Dollars, *for value received, at the First National Bank.*

E. R. Smith.

NOTE 2.

$216 \tfrac{15}{100}.   Albany, N. Y., July 10, 1896.

One year after date, *for value received, I* promise to pay *Asa G. Tucker* or *bearer, Two hundred sixteen* and $\tfrac{15}{100}$ *Dollars, with interest.*

John T. Brown.

NOTE.—A note may be payable on a given day; as, On March 15 after date, I promise to pay, etc. A note may be payable on demand; as, On demand I promise to pay, etc.

290. A note made payable to the payee only is called a non-negotiable note.

NOTE.—When a note is payable in a State in which three days of grace are allowed, maturity is three days after the expiration of the interval named in the note.

291. 1. Write a negotiable note, bearing interest.

Indorse it with payee's name, and find the amount at maturity.

To whom must the maker pay the money?

SIMPLE INTEREST.

2. Write a non-negotiable note payable on a specified date, and find the amount due at maturity.

3. Write a negotiable note, and find the amount of it.

4. Write a non-interest-bearing demand note.

5. Write a non-negotiable interest-bearing note.

6. Write a 6% note, dated June 15, 1895, payable in 1 year without interest, with yourself as payee, and your teacher as maker, and find the amount of it to the present time.

7. Write a negotiable note, using the following :

Date, Jan. 16, 1894 ; Time, 6 months; Face, $1684.96; Payee, Andrew Jackson ; Maker, Silas Wright; Interest at 6%. Indorse it, showing that the maker has transferred it to another. What is the amount of the note, if paid in full Nov. 11, 1894 ?

292. A note should contain :
1. The face in figures at the left upper corner.
2. The place and date at the right upper corner.
3. The time of payment.
4. The words " Value Received."
5. The face written in words in the body of the note.
6. The place at which it is payable.
7. The words " with interest," if agreed upon.

293. A note is said to mature on the day on which it is due.

294. A note that does not contain the words *"with interest"* bears interest from maturity, if not paid at that time.

When does interest begin in Note 1 ? In Note 2 ?

295. In many of the States the maker is allowed three days (called Days of Grace) in which to pay a note, after the time named in the note has expired. In these States, the date of maturity falls on the last day of grace.

Days of grace are not allowed in California, Connecticut,

District of Columbia, Idaho, Illinois, Maryland, Massachusetts, Montana, New Jersey, New York, North Dakota, Ohio, Oregon, Pennsylvania, Utah, Vermont, and Wisconsin.

NOTE. — When the holder of a note transfers it to another, he is usually required to *indorse* it, i.e., to write his name across the back. This is required as an order to the maker to pay the money, when due, to the new holder. An indorser is also responsible for the payment of a note in case the maker fails to pay it when due.

PARTIAL PAYMENTS.

296. Payments in part of a note or other debt are **Partial Payments**.

The Supreme Court of the United States has adopted the following rule for finding the amount due on a note after partial payments have been made.

UNITED STATES RULE.

Find the amount of the principal to the time when the payment or sum of the payments equals or exceeds the interest then due.

Deduct from this amount the payment or payments.

Treat the remainder as a new principal, and so proceed until the date of settlement.

NOTE. — When a partial payment of a note or other contract is made, the holder writes upon the back of it the sum paid, with the date of payment. Sums so written are called indorsements. The common form of indorsement is as follows:

Received on the within,
July 16, 1896, $..........

1. A note was given Jan. 1, 1892, and settled July 13, 1894. The following payments were indorsed upon it: May 25, 1892, $250; Jan. 25, 1893, $45; April 7, 1893, $375; July 13, 1893, $750. How much was due on the day of settlement, interest at 6%?

First write the note, and properly indorse the payments upon the back of it.

SIMPLE INTEREST. 157

YR.	MO.	DA.	PAYMENTS.		
1892	5	25	$250	$1820 Principal.	
1892	1	1		.024	
	4	24		$43.68	
				1820.00	
	.024			$1863.68	1st Amount.
				250.00	1st Payment.
~~1893~~	~~1~~	~~25~~	$45	~~$1613.68 New Principal.~~	
~~1892~~	~~5~~	~~25~~		~~.04~~	
	~~8~~	~~0~~		~~$64.55 Interest exceeds Payment.~~	
	~~.04~~				
1893	4	7	$375	$1613.68	
1892	5	25		.052	
	10	12		$83.91	
				1613.68	
	.052			$1697.59	Amount.
			$420	420.00	Sum of 2d and 3d Payments.
1893	7	13	$750	$1277.59	New Principal.
1893	4	7		.016	
	3	6		20.44	
				1277.59	
	.016			$1298.03	Amount.
				750.00	4th Payment.
1894	7	13	Settled.	$548.03	New Principal.
1893	7	13		.06	
	1	0	0	$28.88	
				548.03	
	.06			$576.91	Ans.

NOTE. — The $45 payment, being less than the interest ($64.55), is not deducted from the amount of the second principal ($1678.23). If this were done, and the remainder treated as a new principal, a portion of it ($19.55), being interest, would draw interest, which is not legal. Therefore, interest must be taken on $1613.68 until the date of the next payment (10 mo. 12 da.). The sum of the two payments, being greater than the interest, is subtracted from the amount.

Write in proper form on paper a note for each of the following, indorse the payments, and solve:

2. Date, Jan. 1, 1874, at Syracuse, N.Y. Face, $1000. Interest at 6%. Indorsements: July 7, 1874, $400; Oct. 19, 1874, $300; Dec. 1, 1874, $100. What remains due Jan. 1, 1875?

3. Face, $900. Date, March 1, 1886. Interest at 9%. Indorsements: Aug. 10, 1886, $300; Sept. 1, 1886, $100; Jan. 1, 1887, $50. What was due March 1, 1887?

4. Face, $2000. Date, Jan. 20, 1892. Interest at 6%. Indorsements: May 20, 1892, $100; July 20, 1893, $100; Sept. 10, 1893, $700; Oct. 20, 1894, $75. Settled Oct. 20, 1895. What was due?

5.

$300.   Binghampton, N.Y., *Oct. 12, 1889.*

On demand, for value received, *I* promise to pay ———— *S. D. Cleveland* ———— or order, *Three hundred dollars,* with interest.

J. H. Van Alstyne.

The following payments were made on this note: June 27, 1891, one hundred fifty dollars; Dec. 9, 1892, one hundred fifty dollars. What was due Oct. 9, 1895?

6. On a note for $573.25, at 6%, dated June 10, 1888, were the following indorsements: May 20, 1889, $50; July 10, 1890, $16.50; April 5, 1891, $14.30; July 14, 1892, $250. How much was due Sept. 20, 1893?

7. A note of $850 was dated June 21, 1892, bearing interest at 6%. On this note were the following indorsements: Sept. 15, 1892, $150.90; Nov. 21, 1893, $45; Jan. 15, 1894, $256.88. What remained due June 21, 1894?

8. Find what was due June 1, 1896, on a note for $1928, with $4\frac{1}{2}$% interest, dated Jan. 1, 1891, and bearing the following indorsements: March 1, 1891, $300; Oct. 16, 1893, $40; Feb. 4, 1894, $800; Dec. 16, 1895, $500.

SIMPLE INTEREST.

9. On a note for $832.26 dated Aug. 3, 1889, due in 6 months, the following payments were indorsed: $350, Oct. 5, 1890; and $468.37, May 15, 1892. How much was due Dec. 12, 1893, interest at 7%?

10. Face, $2950. Date, July 1, 1885. Interest, 7%. Indorsements: Oct. 1, 1885, $750; Jan. 15, 1886, $600; July 1, 1886, $900; Dec. 1, 1886, $300; March 1, 1887, $450. What was due July 1, 1887?

MERCHANTS' RULE.

297. When notes and accounts are settled within a year after interest begins, and upon which partial payments have been made, it is customary for business men to make use of the following rule:

Find the amount of the entire debt at date of settlement.
Find the amount of each payment at date of settlement.
Subtract the amount of the payments from the amount of the debt.

1.

$648.   Syracuse, N.Y., *May 1, 1896.*
For value received, I promise to pay ~~~ J. McCarthy & Co., ~~~ or bearer, Six hundred forty-eight Dollars on demand, with interest.   *Chas. E. White.*

Indorsements: June 1, 1896, $150; Aug. 1, 1896, $200; Oct. 1, 1896, $300.
What was due Dec. 1, 1896?

SOLUTION. — $648 in 7 mo. amounts to $670.68
$150 in 6 mo. amounts to $154.50
$200 in 4 mo. amounts to $204.00
$300 in 2 mo. amounts to $303.00   661.50
               $9.18

2. On a note of $1186.48, with interest at 5%, dated April 4, 1890, these payments were indorsed: July 10, 1890, $250; Aug. 4, 1890, $300; Dec. 8, 1890, $150; Jan. 2, 1891, $75. How much was due Feb. 4, 1891?

3. On Oct. 16, 1896, John D. Wilson gives his note for $483.98, with interest at 6%. He pays the note in full, March 28, 1897, having made a payment of $350 on Jan. 28, 1897. How much does it require to settle the note?

COMPOUND INTEREST.

298. **Compound Interest** is interest on unpaid interest, as well as on the principal, at the end of regular interest periods.

NOTE. — Interest is compounded annually, semi-annually, or quarterly, according to agreement.

Compound interest is not authorized by law. It is customary for savings-banks to allow interest on interest when it has been on deposit for a full interest period.

1. Find the compound interest of $350 for 2 years and 6 months at 6%.

SOLUTION. —
$350.00 Principal.
21.00 Interest for 1st year.
$371.00 Amount taken as new principal.
22.26 Interest for 2d year.
$393.26 Amount used as new principal.
11.80 Interest for 6 mo.
$405.06 Amount for 2 yr. 6 mo.
350.00 1st principal.
$55.06 Compound interest for 2 yr. 6 mo.

NOTE. — When the interest is compounded semi-annually, the rate is one-half the annual rate for each period. When quarterly, one-fourth, etc.

When no interest period is mentioned, interest is compounded annually.

2. What is the compound interest of $830 for 3 years at 5 per cent?

3. What is the amount of $650 for 4 years at 4% interest, compounded semi-annually?

SIMPLE INTEREST.

4. What is the compound interest of $365 for 2 yr. 7 mo. 18 da. at 6%, compounded semi-annually?

5. What is the compound interest on $640 for 4 years at 5%?

6. What is the interest, compounded quarterly, on $538.25 for 2 yr. 6 mo., rate 4%?

7. What is the interest, compounded annually, on $683.48 for 4 years at 6%?

8. What is the compound interest on $437.50, 3 yr. 6 mo., at 5%, compounded semi-annually?

REVIEW OF INTEREST

299. 1. What is simple interest? Compound interest? A promissory note? A negotiable note?

2. Define payee, holder, signer or maker.

3. Describe two common methods of computing interest.

4. Prove that, at 6%, 6 cents is the interest on $1 for 1 year.

5. Prove that 5 mills is the interest on $1 for 1 month.

6. Prove that ⅙ mill is the interest on $1 for 1 day.

7. Why is interest not accurate when computed by the 6% method?

8. Find the interest on $50000 for 252 days by the 6% method, then by the exact interest method. Which is more favorable to the payee?

9. When does a note mature?

10. What elements must be given when we find interest? Rate? Time? Principal?

11. How do you find the rate? The time? The principal?

12. What are days of grace?

13. Does the maker of a non-interest-bearing note ever have to pay interest? Explain.

14. What use is made of compound interest?

15. Find the compound interest, then the simple interest, at 6% on $25000 for 5 years, and note the difference.

16. When a note is not paid at maturity, why is it to the holder's advantage to require a new note?

17. What is the effect of a payee's indorsement?

18. When a partial payment is made that does not equal the interest due, why is not the payment subtracted from the amount?

19. Solve a problem in partial payments by both the United States, and the merchants' rule. Which is more favorable to the payer?

20. Find the compound interest on $1420.80 for 1 yr. 9 mo. at 6%, computed semi-annually.

21. Find the amount of a debt of $5672.00 for 4 years at 4% compound interest.

22. Find the interest on $720 at 6% for 2 yr. 8 mo. 22 days.

Find the interest on:

23. $675.20 for 3 yr. 5 mo. at 7%.

24. $754.30 for 1 yr. 4 mo. 15 da. at $5\frac{1}{2}$%.

25. $564.11 for 2 yr. 3 mo. 18 da. at 4%.

26. A county in Missouri owes $85,640. In how many days will the interest at 6% amount to $897.22?

27. Find the interest at 8% on $3960.36 for 9 mo. 20 days.

28. Find the amount of $2536.48 for 1 yr. 3 mo. 18 da. at 7%.

29. The interest on $600 for 3 yr. 6 mo. was $126. What was the rate?

SIMPLE INTEREST.

30. The interest on a note for $460.50 at 5% was $60.44. What was the time?

31. The interest on a certain sum was $96.04, the rate 6%. Find the principal.

32. The amount due on a 6% note due in 1 yr. 5 mo. 1 da. was $135.708. What was the face of the note?

33. Find the exact interest on a note for $600, dated Aug. 5, 1895, and due July 1, 1896, interest at 6%.

34. Find the amount and simple interest of $623.74, one half of which is to be paid in 2 yr. 3 mo. at 4%, the other half to be paid in 3 yr. 5 mo. at 6%.

35. A note for $146.20, dated June 5, 1869, was paid July 11, 1872, with interest at 6 per cent. What was the interest?

36. A man borrowed, Dec. 25, 1877, $137.40 at 6% interest, and kept it until Jan. 15, 1880. What was the interest?

37. Payments were made on a note of $1800 dated Jan. 12, 1891, as follows: March 6, 1891, $300; April 15, 1891, $190; July 3, 1891, $565; Oct. 15, 1891, $700. What was due Dec. 21, 1891, interest at 6%?

38. When must $1600 be put at interest at 6%, so that it will amount to $1800 on Jan. 1, 1898?

39. Find the amount of $375 for 2 yr. 8 mo. 16 da. at 6%.

40. Find the amount at simple interest of $1200 from April 4, 1895, to the present time.

41. A note for $728 is dated Nov. 16, 1894. March 8, 1895, there was paid on it $25. Find the amount due on Jan. 4, 1896, interest at 6%.

42. Find the amount at simple interest of $1184.63 for 1 yr. 4 mo. 17 da. at $4\tfrac{1}{2}$%.

43. Write your own promissory note for $200, with interest, payable in 60 days from to-day. When does it become due? Find the amount due at maturity.

44. Find the exact interest on $843.20 from April 10, 1895, to March 15, 1896, at $4\tfrac{1}{2}\%$.

45. Upon a note for $950, dated Syracuse, N. Y., Jan. 1, 1894, $150 was paid Aug. 16, 1894; $25 March 1, 1896; and $200 April 16, 1896. How much is due to-day?

46. $645 was paid as interest on $2000 for 3 yr. 7 mo. What was the rate?

47. $30 was paid as interest on $600 at 6%. What was the time?

48. A house that cost $5000 was rented for $500, and $100 was paid for annual taxes and repairs. What rate of interest did the investment yield?

49. A person investing a certain sum of money at 6% for 1 yr. 6 mo. found at the end of that time the investment amounted to $545. Find the sum invested.

50. A man bought a horse for $150, paying $70 in cash, and the balance on time at 6%. He paid at the time of settlement $83.60. How much time elapsed before that date?

51. H. C. Harmon loaned $250 for 1 yr. 3 mo. 27 da., which amounted to $269.875 at the time of payment. Find the rate of interest.

52. A person having a certain sum of money invested, and drawing compound interest at 6%, found at the end of 2 yr. 2 mo. that it amounted to $567.418. What was the sum invested?

53. A sum of money was borrowed Jan. 30, 1895, and $419.60 paid in full Nov. 24, 1895. The rate of interest being 6%, how much of this was interest?

54. A man owes $4600 at 7%, and each payment of interest amounts to $161. How often does he pay interest?

TRUE DISCOUNT.

300.. Oral.

1. What will be the amount of $100 at 6% one year from to-day?

2. What is the value to-day of a debt of $106, due in one year, when money is worth 6% interest?

3. How much money paid to-day will cancel a debt of $112, due two years hence, money being worth 6%?

4. What is the present worth of $105, due in one year without interest, when money is worth 5% interest?

5. When money can be loaned at 7%, which is worth the more, $100 at the present time, or a note of $107 without interest, due in one year?

6. What sum should be deducted from a debt of $108, due without interest in one year in consideration of its being paid now, when money can be loaned at 8%?

301. True Discount is a deduction of interest for the payment of a debt before due.

302. The **Present Worth** of a debt due at a future time is a sum which will amount to the debt if put at interest till that time.

The debt is therefore the amount of the present worth for the given time.

303. The true discount is the difference between the debt and its present worth. It is the interest of the present worth for the given time.

7. What is the present worth and the true discount of a debt of $582.40, due in 8 months without interest, when money is worth 6%?

Solution. — $582.40 ÷ $1.04 = $560, present worth.
$582.40 — $560 = $22.40, true discount.

Since $1.04 is the amount of $1 for 8 mo., $582.40 is the amount of as many dollars as $1.04 is contained times in $582.40 = $560.

Rule. — *To find the present worth, divide the debt by the amount of $1 for the given time.*

To find the true discount, subtract the present worth from the debt.

8. What is the present worth and true discount of $400, due in one year, when money is worth 5%?

9. A father wills his two sons $3000 each, to be paid in three years from the time of his death. What is the value of the legacies at the probate of the will, if money is worth 6%?

10. What is the present worth of $450, due in two years at 5%?

11. What is the present worth of $250.51 payable in 8 months, money being worth 6%?

12. Which is better, to buy flour for $5 cash, or for $5.25 on 6 months' time, when money can be borrowed at 5%?

13. Find the present worth of $750 for 6 months, money being worth 6%.

14. What is the present worth of $600, due in 1 year without interest?

15. Write the note which would be given for the above debt.

16. A man wishing to buy a house and lot has his choice between paying $5400 in cash, or $4000 in cash and $1700 in two years. With money at 6%, which is the most advantageous for him?

17. Which would be more profitable, and how much, to pay $4000 cash for a house, or $4374.93 in 3 yr. 6mo., money being worth 7%?

18. I can sell my house for $2800 cash, or $3000 and wait 6 months without interest. I choose the latter; do I gain or lose, and how much, money being worth 6%?

19. What is the present worth of a debt of $385.31, due in 5 months, 15 days, at 6%?

BANK DISCOUNT.

304. When the holder of a negotiable note wishes the money before it becomes due, he may take it to a commercial bank; and if the banker is satisfied that the parties to the note are responsible, he will pay the holder the amount due after deducting the discount. By this act the bank becomes the holder of the note, and at its maturity the maker must pay to the bank instead of to the payee.

305. The **Maturity Value** of a note is the amount due at maturity.

The **Bank Discount** is the simple interest on the maturity value, reckoned from the day of discount to the day of maturity.

306. The maturity value less the bank discount is called **Proceeds, or Avails**. The time from the day of discount to the day of maturity is called the **Term of Discount.**

307. The maturity value of a note, not bearing interest is the face, and the maturity value of an interest-bearing note is the face plus the interest.

NOTE 1. — Only short-time notes are discounted at banks, usually not exceeding 4 months.

NOTE 2. — Banks generally require that the paper which they discount be made payable at some bank.

NOTE 3. — Banks usually reckon discount for the *exact number* of days in the term of discount, although the time in a note may be expressed in months. Banks usually regard the year as 360 days.

NOTE 4. — In States having days of grace, the day of maturity is the last day of grace.

NOTE 5. — If the day of maturity falls on Sunday or a legal holiday, the preceding day is the day of maturity in most States. In some States, however, the note does not fall due until the day following.

NOTE 6. — When a note is discounted at date, the term of discount is the time of the note (+ 3 days of grace in States having days of grace).

Unless otherwise stated, a note is to be discounted at date.

1.

$325 $\tfrac{24}{100}$ Syracuse, N. Y., *Jan 15, 1896*

Two months after date, for value received, *I* promise to pay _____ *Richard Turner;* _____ or order, *Three hundred twenty-five* and $\tfrac{24}{100}$ Dollars, at the Third National Bank.

Henry P. Warner

If this note was discounted at 6% at a bank on the day it was made, how much did the bank deduct? How much were the proceeds?

SOLUTION. — The term of discount is from Jan. 15 to Mar. 15 =

JAN.   FEB.   MAR.
16 da. + 29 da. + 15 da. = 60 da.

The bank discount is the interest of 325\tfrac{24}{100}$ for 60 da., at 6% = $3.25; the proceeds = $325.24 − $3.25 = $321.99.

2. Copy the above note, and properly indorse the payee's name across the back.

3. What would be the bank discount and proceeds of the above note if it contained the words "with interest."

SOLUTION. —
   Maturity value = face + interest, or $325.24 + $3.25 = $328.49.
   The bank discount = 6% of $328.49 for 60 da. = $3.28.
   The proceeds = $328.49 − $3.28 = $325.21.

4.

$387 50/100.   Boston, Mass., June 27th, 1896.

Three months after date, I promise to pay to the order of_____ James G. Rogers, _____
Three hundred eighty seven and 50/100 Dollars, value received, at the First National Bank, with interest at 5%.

George Price.

Discounted July 27 at 5%.
Day of maturity, Sept. 27.

SOLUTION. —
   Maturity value = face + interest for 90 da., at 5% = $392.34.
   Discount at 5% from July 27 to Sept. 27, 62 da. = $3.38.
   Proceeds $392.34 − $3.38 = $388.96.

5.

$648.15.   Buffalo, N. Y., Jan. 31, 1895.

One month after date, I promise to pay to the order of_____ James B. Strong, _____
Six hundred forty-eight and 15/100 Dollars, at the Shoe and Leather Bank, value received, with interest at 6%.

Discounted at date at 6%.
The above note is interest-bearing, therefore the discount must be computed on the amount at maturity.

6.

$3000.   Detroit, Mich., *Oct. 1, 1896.*

Ninety days after date, for value received, *I* promise to pay _____ *Jerome K. Nixon,* _____ or order, *Three thousand dollars,* at the First National Bank.

Leroy C. Bondy.

Discounted at date at 5%.

7.

$438.29.   St. Louis, Mo., *Feb. 7, 1893.*

Two months after date, for value received, *I* promise to pay _____ *Hugh Thompson,* _____ or order, *Four hundred thirty-eight and* $\frac{29}{100}$ *Dollars*, at the Chemical National Bank.

R. H. Winthrop.

Discounted March 10 at 6%.

8.

$789 $\frac{50}{100}$.   Cleveland, O., *Mar. 4, 1896.*

Four months after date, *I* promise to pay to the order of _____ *W. W. Woodford* _____ *Seven hundred eighty-nine and* $\frac{50}{100}$ *Dollars*, Value received, at the City Bank.

Otis R. Young.

Discounted May 4 at 6%.

9.

$4920.   Brooklyn, N. Y., Apr. 3, 1890.

Ninety days after date, for value received, I promise to pay to the order of ~~~~~ *Dewitt Long* ~~~~~ *Four thousand nine hundred and twenty* Dollars, at the Merchant's Bank, with interest.

Elizabeth R. Prentiss

Discounted at date at 6%.

10.

$1312.   Boston, Mass., May 1, 1892.

Sixty days after date, for value received, I promise to pay to the order of ~~~~~ *Edgar N. Wilson* ~~~~~ *One thousand three hundred and twelve* Dollars, at the First National Bank.   *Frank L. Barker*

Discounted May 10 at 4%.

11.

$2142.84.   Albany, N. Y., Dec. 15, 1896.

Five months after date, for value received, I promise to pay ~~~~~ *Charles R. Skinner* ~~~~~ or order, *Two thousand one hundred forty-two* $\frac{84}{100}$ Dollars, with interest, at the Park Bank.   *R. H. Dixon.*

Discounted Feb. 27, 1897, at 6%.

12.

$2000.   San Francisco, Cal., June 10, 1896.

Three months after date, for value received, I promise to pay ~~~~Seymour G. Wilcox~~~~ or order, Two thousand Dollars, at the Citizen's Bank.

P. J. Reed.

Discounted July 10 at 8%.

13.

$2500   Syracuse, N. Y., July 6, 1896.

Two months after date, I promise to pay to the order of ~~~~Robert Beecher,~~~~ Two thousand five hundred Dollars, at the Third National Bank.

John Q. Adams.

Discounted at 6% at date.

14. A note for $135 is given for 90 days, and discounted the day it is given at 6%. What are the proceeds?

15. William Johnson gave John Doe a note payable to the Binghamton Trust Co., time 60 days, amount $204.60. Write this note. After 20 days Doe put the note in the bank. What are the proceeds of the note?

In the following problems, write the notes in full, and properly indorse them, using any names for payer and payee.

16. Find the bank discount of $400 for 3 months at 8%.

17. What are the proceeds of $250, with interest at 6%, discounted at bank for 60 days at 6%?

DISCOUNT. 173

18. What will be the proceeds of a note for $175 drawn at 4 mo., with interest at 6%, if the bank discount is 10% per annum?

19. On the first day of January, 1896, a farmer gave his note at 90 da. for $525, with interest at 6%. When did the note become due? and what were the proceeds of the note if discounted at a bank at 1% a month on the tenth day of February?

308. To find the Face of a note, when the Proceeds, Time, and Rate are known.

1. What must be the face of a 60-day note, without grace, which after being discounted at 6% will give $500 as proceeds?

SOLUTION. —
  The bank discount of $1 at 6% for 60 da. = $.01.
  The proceeds of $1 = $1.00 − $.01 = $.99.

Since $.99 is the proceeds of $1, $500 must be the proceeds of as many dollars as $.99 is contained times in $500 = 505.05+   *Ans.*

Therefore, $505.05 must be the face of a 60-day note which will give $500 as proceeds after being discounted at 6%.

Rule. — Divide the proceeds by the proceeds of $1.

2. A person must use $250 to-day. For how much must he make a bank note for three months that will give $250 proceeds, without grace?

3. What must be the face of a 60-day note, payable at a Boston bank, upon which I can realize $350 after it is discounted at 6%?

4. If you buy goods for $1200 cash, how large a note payable in 90 days, at 6% bank discount, must you make that the proceeds shall pay for the goods? Without grace.

5. Find the face of a 60-day note that will yield $800 when discounted at bank at 7%, with grace.

6. How large a note must I make at a bank for 30 days to pay a debt of $475, without grace?

7. Wishing to borrow $494.90 at a Syracuse bank, for what must I make my note at 60 da., with interest at 6%, in order to obtain this amount? Discount at $\frac{1}{2}$% a month.

8. The proceeds of a Buffalo note at 60 da., when discounted at a bank at 6% per annum, is $742.50. What is the face?

REVIEW OF DISCOUNT.

309. 1. Define Discount; True Discount; Bank Discount; Proceeds; Present Worth.

2. How is the present worth found? The true discount? The bank discount? The proceeds?

3. What is the term of discount? and how is it found? The day of discount? and how found?

4. How is the bank discount of an interest-bearing note found?

5. How do you find the face of a note when the proceeds, time, and rate are given?

6. When does a Rochester, N.Y., note mature if given for 1 month from Jan. 31?

7. State a point of difference between true discount and bank discount.

8. What kind of notes only can be discounted at banks?

9. Bought a city lot, and agreed to pay $546.94 at the end of 2 yr. 6 mo., without interest. Receiving some money unexpectedly after 6 months, I wish to pay cash. How much ought I to pay, money being worth 6%?

10. What is the present value and true discount of $973.52, due in 1 yr. 7 mo. 24 da. hence, without interest, money being worth 8%?

DISCOUNT.

11. A man has an offer of $2000 cash for his house, or $2100 payable in 8 months. If money is worth 8%, which is the better? and how much?

12. Find the discount and proceeds of a note for $13500 payable at a bank in 90 days after date without grace, discounted at 5%.

13. For what sum must C. F. Norton draw his note on a Binghamton bank, that when it is discounted at 4% for 60 days he will have $800?

14. A man owes me $2540, due in 2 years, 3 months, without interest. If he pays it at once, what discount should I allow him?

15. Find the discount and proceeds of a note on a Brooklyn bank for $350, given May 12, 1896, for 4 months, and discounted at 6%, July 15.

16.

$860.   Syracuse, N. Y., *May 5, 1896.*

Three months after date, for value received, *I* promise to pay _____ *R. B. White* _____ or order, *Eight hundred sixty* Dollars, with interest, at the Salt Springs National Bank.   *H. E. Barrett*

Discounted June 11 at 6%.

17. For what sum must I draw a 4 months' note so that the proceeds will be $800, discounted without grace at 6%?

18. I sell my horse for $216, and take a note due in 6 months without interest. If money is worth 6% per annum, what is the present value of my note?

19. For what sum must I give my note for 60 days at a bank in order to receive $650 proceeds, money being worth 8%?

20. Find the face of a note, discounted for $2558.40 at 8%, for a term of 72 days, without grace.

STOCKS AND BONDS.

310. Many kinds of business require so much capital that several persons must unite to raise the necessary amount.

311. The Capital to be raised is divided into **Shares**, usually of $100 each.

Shares are then sold until the required amount is raised.

Each purchaser of shares is a **Stockholder,** and receives a **Certificate of Stock,** which shows the number of shares purchased, and their value.

This value is called the Par or Nominal Value.

312. The Market Value of stocks is the price for which they are sold.

313. The value of stocks depends upon the profitableness of the business. When the business is very profitable. the shares are worth more than par; they are then above par, or at a Premium.

When the business is unprofitable, the shares are not worth their par value. They are then below par, or at a Discount.

314. 1. The capital of a company is $100000. Into how many shares of $100 each can this be divided?

2. A stockholder owns 25 shares of stock. How many dollars of stock has he?

STOCKS AND BONDS. 177

3. If at the end of a year there has been a net profit of $10,000, what per cent profit has been made?

$10000 is what % of $100000?

The profits are divided among the stockholders, and are called Dividends.

NOTE 1. — Dividends are usually declared semi-annually or quarterly.

NOTE 2. — When a 10% dividend is declared, each stockholder receives 10% of the par value of his shares.

4. What will be A's dividend if he owns 35 shares?

When there is a loss, each stockholder is required to pay his share of the loss. This is called an Assessment.

5. What would be A's assessment to meet a 2% loss?

A person who buys or sells stocks for others is called a Stock-broker, and his commission is called Brokerage.

NOTE 1. — Brokerage is usually ¼% or ⅛% of the par value.

NOTE 2. — In all stock transactions, dividends, assessments, brokerage, premium, and discount are computed on the par value.

NOTE 3. — Shares are sometimes issued at $200, $250, $50, $25, or $10 each, but unless otherwise stated $100 is considered the par value of a share.

6. What is the market value of 10 shares of bank stock, when sold at par?

7. What is the market value of 50 shares of railroad stock, at 10% premium?

SOLUTION. — The market value of 1 share is $100 + $10 = $110.
    The market value of 50 shares is 50 times $110.

8. What is the market value of 18 shares of mining stock at 15% below par?

SOLUTION. — The value of 1 share is $100 − $15 = $85.
    The value of 18 shares is 18 times $85.

Stock Quotations are the published prices of stocks.

When railroad stock is quoted at 108, it means that it sells for 8% above par in the stock-market.

When it is quoted at 92, it is selling at 8% below par.

9. If I buy stock at 98 and sell it at 101, what gain do I make on 10 shares?

NOTE. — Stock at 98 means $98 for a $100 share, and stock at 101 means $101 for a $100 share.

10. When stock is quoted at 85, what is the value of a share? What is the value of 1 dollar of stock?

11. What must I pay for 10 shares of stock at 95, if I pay the broker ¼% for doing the business?

SOLUTION. — Cost of 1 share = $95 + Brokerage $¼ = $95¼ or $95.25.
$95.25 × 10 = $952.50. *Ans.*

12. If I sell 10 shares of stock at 110, and pay the broker ¼%, what do I receive?

SOLUTION. — 1 share brings $110 − $¼ = $109¾, or $109.75.
10 shares bring 10 times $109.75 = $1097.50. *Ans.*

13. A man invested $4500 in street railway stock at 10% discount. How many shares did he purchase?

14. If I invest $2100 in bank stock at 105, how many shares do I purchase?

15. A capitalist bought 80 shares railroad stock at 87½, and 60 shares mining stock at 114. Find the cost.

16. $18200 will purchase how many shares of stock selling at 140?

17. A stock company declares a dividend of 2½%. What does A receive, who owns 1500 shares of $10 each?

18. How much is gained on 50 shares of railroad stock purchased at 98 and sold at 102?

19. Bought stock at a discount of 2%, and sold it at a discount of 3%. Did I gain or lose? and how much on 20 shares?

BONDS.

315. To meet extraordinary expenses, governments, States, cities, villages, counties, towns, and incorporated companies sometimes borrow money. The securities given by such corporations are called **Bonds**.

Bonds bear a fixed rate of interest, payable annually, semi-annually, or quarterly. They are bought and sold in the same manner as stocks.

Bonds are known by the rate of interest they bear: Virginia 6's are bonds of the State of Virginia, bearing 6%; U. S. 4's of '97 are U. S. bonds bearing 4% interest, and maturing in 1897.

316. A **Coupon** is an interest certificate attached to a bond. At the expiration of any interest period, the coupon is cut off and used in collecting the interest, being worth the amount of interest due on the bond for a specified period.

317. 20. What will be the cost, including brokerage at $\frac{1}{4}$%, of 200 shares of C., B., and Q. R.R. bought at $67\frac{7}{8}$?

SOLUTION. — Cost of 1 share = $67\frac{7}{8}$ + $\frac{1}{4}$ = $68\frac{1}{8}$. Cost of 200 shares = 200 × $68\frac{1}{8}$.

21. How much, including brokerage at $\frac{1}{8}$%, must be paid for $5000 of U. S. 4's at $110\frac{7}{8}$?

SOLUTION. — $1 of bonds costs 1.10\frac{7}{8}$ + .00$\frac{1}{8}$ = $1.11. $5000 worth will cost 5000 times $1.11.

22. What must I pay for $8275 of stock at 10% discount?

23. What is the cost, including broker's commission of $\frac{1}{4}$%, of 150 shares of railroad stock bought at $89\frac{1}{4}$?

24. I buy stocks at 5% discount, and sell at 5% premium; what per cent profit do I make on the investment?

25. March 10, 1896, Western Union Telegraph stock was quoted at 84⅞. How many shares could be bought for $1020, brokerage ⅛ per cent?

SOLUTION. — Cost of 1 share, $84⅞ + ⅛ = $85. As many shares can be purchased as $85 is contained in $1020.

26. How many shares of stock at 10% premium can be purchased for $2200?

27. I invested $5100 in N.Y. and N.H. railroad stock at 170. How many shares did I purchase?

28. If I invest $42400 in 5% bonds at 106, what is my yearly income?

SOLUTION. — $42400 ÷ $1.06 = $40000, par value. How much is 5% of $40000?

29. If I invest $21008 in 5% bonds at 104, what will be my annual income?

30. What will be my yearly income if I invest $11100 in 5% at 92, brokerage ½%?

31. A man invests $9500 in Virginia 6's at 94¾, brokerage ¼%. What is his quarterly income?

32. What will be my annual income if I invest $5050 in 4% water bonds at 1% premium?

33. What is my dividend on 80 shares of electric-light stock, when a 5% dividend has been declared?

34. What sum must be invested in Chicago 5's at 92 to yield an income of $600, brokerage ½%?

SOLUTION. — $600 ÷ .05 = $12000, par value. How much is 92½% of $12000?

35. How much must I invest in 4% bonds at 8% premium, to secure an annual income of $200?

STOCKS AND BONDS. 181

36. How much must be invested in city 3½'s at 8% discount, to secure an income of $350 ?

37. How much telegraph stock must I sell at 11½% discount, brokerage ½%, to realize $8800 ?

38. I invested through a broker $5450 in stock at 1.08½, brokerage ½%. How much did I purchase ?

39. I sell through a broker enough stock at 4⅞% premium to realize $10475, brokerage ½%. How much do I sell ?

40. What rate of interest do I receive on my investment if I buy 7% stock at 112 ?

SOLUTION. — Each share of stock costs $112, and yields $7 interest. $7 is what per cent of $112 ?

41. Stock yielding 7% annually is bought at 111¼. What annual rate of income will it yield on the investment ?

42. What rate will 6% bonds pay on the investment if bought at 112 ?

43. What is the rate on Syracuse 4's at a premium of 3¼% ?

44. What is the rate of income on 6's at 90, no brokerage ?

45. Which is the better investment, 6's at par, or 5's at a discount of 12½% ?

46. How much must I pay for 5% stock to secure annually 7% on my investment ?

SOLUTION. — 1 share of 5% stock yields $5 interest annually; this $5 is 7% of the cost of 1 share. Therefore the question is, $5 is 7% of what ?

47. At what price must 5% stock be purchased so that it will yield 4% on the investment ?

48. How much must I pay for 5's to make my investment yield 6% ?

49. What must I pay for city 6's that my investment may yield 8% annually ?

50. How much must I pay for 1 share of 3% stock, that the dividend may be 4% of the purchase price?

51. How much will be my income if I invest $2300 in 4% bonds at 115?

SOLUTION. — 2300 ÷ $1.15 = $2000 par value. How much is 4% of 2000?

52. What sum invested in Tennessee 6's at 85 will yield an annual income of $1800?

53. How much money must I invest in 6% stock at 80 to secure an annual income of $3186?

54. I want an income of $1500. How much shall I invest in 5% stocks at 25% premium to secure that amount?

55. How much must a man invest in a 5% stock at 120 to yield him an annual income of $2500?

MISCELLANEOUS.

318. 1. At what premium should an 8% stock sell to yield a 6% income?

2. A man bought stock at $3\frac{1}{2}$% discount and sold it at 2% premium, paying a brokerage of $\frac{1}{4}$% in both cases. His net profit was $680. How much money did he invest?

3. A man invested his money in 6% railroad stocks, and received $300 semi-annually. What was the sum invested?

4. Which is the better investment, and how much, a 4% stock bought at 85, or a 6% stock bought at 120?

5. What rate on the investment do 7% stocks pay when bought at a premium of 8%?

6. What sum must be invested in U. S. 6% bonds to yield an income of $1000?

7. What sum must be invested in U. S. 6's at $92\frac{1}{2}$ per share to yield a quarterly dividend of $300?

8. At what price should 8% bonds be bought to make the income from the investment equivalent to that from 6% bonds at par?

9. Which is the better investment, 4% bonds at 86, or 6% bonds at 105?

10. How much must I pay for a 4% stock that the investment may yield me 6%? For a 7% stock that the investment will yield 5%?

11. If 25 shares of stock paying 8% are sold at 150, and the proceeds loaned at 6%, will the income be increased or diminished? and how much?

12. Bought bonds at 125 and sold them at 110, thereby losing $600. How many $1000 bonds did I buy?

13. How many dollars of stock can I buy for $105000 if stock is quoted at 120? How many shares? What per cent do I receive on my investment if the stock bears 6%?

14. What is the cost of 200 shares of D., L., & W. R.R. at $162\frac{1}{2}$? If it pays a quarterly dividend of 2%, what is the yearly income from this investment? What rate does it pay on the investment?

15. B invests $1680 in a stock selling at 112. What does he receive from a dividend of 4%?

16. An estate derives an annual income of $3600 from stock that pays $7\frac{1}{2}$%. How many $25 shares does the estate own?

AVERAGE OF PAYMENTS.

319. 1. The use of $5 for 2 mo. equals the use of $1 for how many months?

2. The use of $10 for 6 mo. will balance the use of $5 for how many months?

SOLUTION. — The use of $10 for 6 mo. = the use of $1 for 60 mo.
The use of $1 for 60 mo. = the use of $5 for $\frac{1}{5}$ of 60 mo. = 12 mo.

3. How long may $20 be kept to balance the use of $5 for 20 months? $50 for 10 mo.?

4. A credit of $10 for 8 mo. equals a credit of $20 for how many months?

5. The interest of $500 for 1 year equals the interest of $100 for how long? Prove this.

6. I pay a debt of $20 four months before it is due. How long after it is due should my creditor allow a debt of $40 to remain unpaid?

A person owing two debts due at different times may pay both at an intermediate time without loss to himself or his creditor, by paying one of them before it is due and the other an equivalent time after it is due.

320. The process of finding the time when several debts due at different times can be equitably discharged at one payment is called **Average of Payments**.

321. The date of such payment is called the **Average Time**, and the time to elapse before the payment is made is called the **Average Term of Credit**.

NOTE. — The time to elapse before any debt becomes due is called a Term of Credit.

322. When the terms of credit begin at the same date.

1. On Jan. 8, A bought goods on the following conditions:

$300 due in 2 months.
$200 due in 4 months.
$100 due in 6 months.

How long after Jan. 8 may the debt be equitably discharged at one payment?

SOLUTION. —

A credit of $300 for 2 mo. = a credit of $1 for 600 mo.
A credit of 200 for 4 mo. = a credit of $1 for 800 mo.
A credit of 100 for 6 mo. = a credit of $1 for 600 mo.
   A is entitled to a credit of $1 for 2000 mo.

AVERAGE OF PAYMENTS. 185

A credit of $1 for 2000 mo. = a credit of $600 for $\frac{1}{600}$ of 2000 mo., or $3\frac{1}{3}$ mo. = 3 mo. 10 da., average term of credit.
Jan. 8 + 3 mo. 10 d. = April 18, equated time. *Ans.*

Short method.

$$2 \text{ mo.} \times 300 = 600 \text{ mo.}$$
$$4 \text{ mo.} \times 200 = 800 \text{ mo.}$$
$$6 \text{ mo.} \times 100 = 600 \text{ mo.}$$
$$600 \;\big/\; 2000 \text{ mo.}$$
$$3\frac{1}{3} \text{ mo.} = 3 \text{ mo. 10 da.}$$
$$\text{Jan. 8} + 3 \text{ mo. 10 da.} = \text{April 18.}$$

NOTE. — One-half a day or more is called another day. Less than $\frac{1}{2}$ day, not counted.
Call 50¢ or more $1.00. Less than 50¢, not counted.

Rule. — *Multiply each debt by its term of credit. The sum of the products divided by the sum of the debts will be the average term of credit.*

2. Gates Thalheimer sold a bill of goods on the following terms: $325 due in 60 days, $175 due in 90 days, and $185 due in 4 months. What is the average term of credit? and on what day may the entire debt be paid without loss to either party?

3. A merchant bought $1000 worth of goods, and agreed to pay for them as follows: $100 cash; $300 in 3 mo.; $250 in 4 mo.; and the balance in 5 mo. In what time could he equitably pay the entire amount?

4. On the first day of April, 1895, a man gave 3 notes, one for $250 due in 30 da., one for $375 due in 40 da., and one for $425 due in 60 da. What is the average term of payment? and when could they have all been paid at once?

5. D. McCarthy & Co. sold goods amounting to $4000, payable as follows: $\frac{1}{4}$ in three months, $\frac{1}{4}$ in 4 months, and the balance in 5 months. What was the average term of credit?

6. A merchant sold goods on the following terms: ⅓ payable in 2½ months, ¼ in 3½ months, ⅓ in 5½ months, and the balance in 6 months. What was the average term of credit?

7. Equate the following payments: $580.75 due in 30 days, $650.25 due in 60 days, $450.36 due in 90 days, and $600 due in 5 months.

8. On the 1st of May a merchant bought goods amounting to $1500, agreeing to pay for them as follows: $521.35 on the 10th of June, $398.84 on the 16th of July, $199.60 on the 15th of August, and the balance on the 1st of September. Upon what date can he pay the whole amount?

9. Jacob Amos sold a bill of flour amounting to $2500, payable as follows: $500 due in 4 months, $600 due in 5 months, and the balance due in 6 months. What was the equated time?

10. A purchased a farm for $3000, agreeing to pay for it as follows: $500 cash, $600 in 5 months, $1000 in 8 months, and $400 in 1 year. He decides to give a note for the whole amount. When was the balance to be paid?

323. When the terms of credit begin at different dates.

1. A purchased goods of Dey Bros. & Co., as follows:

   Jan. 8, 1895. $200 on 2 months' credit.
   Feb. 16, 1895. $400 on 3 months' credit.
   April 4, 1895. $300 on 4 months' credit.

Find the average time.

NOTE. — First find the date when each item is due.
   $200 due Mar. 8. 200
   400 due May 16. 400 × 69 da. = 27600
   300 due Aug. 4. 300 × 149 da. = 44700
          900   72300

AVERAGE OF PAYMENTS. 187

72300 ÷ 900 = 80⅓ da. = 80 da.
March 8 + 80 da. = June 27, average time.

The first debt is due March 8, and the last Aug. 4. The average time, therefore, will be between these dates.

$200 due March 8 has no longer time to run.
$400 due May 16 has 69 days after March 8.
$300 due Aug. 4 has 149 days after March 8.

A is therefore entitled to a credit of $1 for 72300 da. after March 8, which is equal to a credit of $900 for 80 da. after March 8.

Rule. — *Find the date on which each debt becomes due, and using the earliest of these as a standard date, reckon the time to each of the others.*

Multiply each debt by its time, and divide the sum of the products by the sum of the debts.

The quotient will be the average term of credit, which add to the standard date to find the average time.

2. Four notes are due as follows: March 4, $165; April 15, $325.50; May 9, $94; June 6, $465. What is the average date of payment?

3. A retail dealer bought the following bills of goods on 4 months' credit: April 4, $480; April 26, $185.65; June 1, $480.16; July 6, $196. What is the average time for payment?

4. Bought goods as follows: Jan. 1, $250 at 3 mo.; Feb. 1, $500 at 4 mo.; March 11, $106 at 60 da. What is the average date of payment?

5. Mr. B owes $1000, due in 5 months; in 2 months he pays $600. How long after the expiration of the 5 months should the remainder be paid?

SOLUTION. — $600 has been paid 3 months before due, which equals a credit of $1 for 1800 months. He is entitled to a like credit for $400 after it is due. $\frac{1}{400}$ of 1800 mo. = 4½ months. *Ans.*

6. A lady purchased a piano for $500 on 6 months' credit. If she pays $200 cash, how long after the expiration of the 6 months should the balance be allowed to run?

7. May 1, 1896, a man buys a store and fixtures for $2650, giving his note payable in 6 months without interest. June 15, he pays $500; Aug. 1, $750. When should the balance be paid?

8. G. L. Hoyt purchased goods of Mann & Hunter to the amount of $3000: $1200 to be paid June 2, 1896; $600 to be paid July 5, 1896; $200 to be paid Aug. 15, 1896. The balance will become due Aug. 30, 1896. At what date must a note payable in 3 mo. be drawn that it may become due at the average date?

QUESTIONS.

324. 1. Define discount; present worth; true discount. Tell how to find present worth and true discount.

2. Define bank discount; proceeds; day of maturity; term of discount.

Tell how to find bank discount and proceeds.

Tell how to find face of note when proceeds, time, and rate are given.

3. What is a stock company? What are stocks? Bonds? Shares?

4. Define par value; market value.

5. What is a stock certificate?

6. Define dividend; assessment.

7. Upon what are premium, brokerage, dividends, and assessments reckoned?

8. What is the average of payments? Equated time? Average term of credit?

RATIO AND PROPORTION.

325. Oral.

1. 5 bears what relation to 10 ? *Ans.* 5 is $\frac{1}{2}$ of 10.
2. 10 bears what relation to 5 ? *Ans.* 10 is 2 times 5.
3. What part of 16 is 4 ?
4. How does $7 compare with $14 ?
5. John has 20¢ and Mary 5¢. What is the relation of John's money to Mary's ? Of Mary's money to John's ?
6. What is the relation of 15 to 3 ? Of $8 to $16 ? Of 28 men to 7 men ? Of 2 bushels to 2 pecks ?

326. Ratio is the relation between two like numbers. It is found by dividing one by the other; thus:

The ratio of 4 to 8 is $4 \div 8 = \frac{1}{2}$.

The sign of ratio is (:). It is the division sign with the line omitted.

The ratio of 6 to 3 is expressed thus, 6 : 3. It may also be expressed fractionally, thus, $\frac{6}{3}$.

327. The **Terms** of a ratio are the two numbers compared.

The first term of a ratio is the **Antecedent,** and the second the **Consequent.**

In the ratio 6 : 12, 6 is the antecedent, and 12 the consequent.

328. A ratio formed by dividing the consequent by the antecedent is an **Inverse ratio.**

$12 \div 6$ is the inverse ratio of 6 : 12.

329. The two terms of a ratio taken together form a **Couplet.**

330. Two or more couplets taken together form a **Compound ratio.**

$\left.\begin{array}{l}3:6\\8:5\\4:5\end{array}\right\} = 96:150$ A compound ratio may be changed to a simple ratio by taking the product of the antecedents for a new antecedent, and the product of the consequent for a new consequent.

Antecedent ÷ Consequent = Ratio.
Therefore, Ratio ÷ Antecedent = Consequent;
and, Ratio × Consequent = Antecedent.

Multiplying or dividing both terms of a ratio by the same number does not change the ratio.

The ratio $12:6 = 2$.
The ratio $3 \times 12 : 3 \times 6 = 2$.
The ratio $12 \div 3 : 6 \div 3 = 2$.

Find the ratio of:

7. $56:7$.
8. $20:300$.
9. $\$55:\$330$.
10. What is the ratio of $\frac{9}{10}$ to $\frac{3}{10}$?
11. 3 bu. : 3 pk.
12. $\frac{1}{2}:4$.
13. $12:\frac{1}{4}$.
14. $2\frac{1}{2}:16$.
15. $\frac{1}{2}:\frac{2}{3}$.
16. $3\frac{2}{5}:5\frac{2}{5}$.

Note. — Fractions with a common denominator have the same ratio as their numerators. Prove this in Ex. 10, by multiplying both terms by 10.

17. $\frac{8}{17}:\frac{16}{17}=?$ $\frac{28}{75}:\frac{7}{75}=?$ $\frac{14}{11}:\frac{30}{11}=?$
18. $\frac{3}{4}:\frac{2}{3}=?$ $\frac{3}{7}:\frac{5}{8}=?$ $\frac{1}{4}:\frac{3}{5}=?$ $\frac{3}{4}:\frac{5}{6}=?$
19. Find the inverse ratio of 75 to 25. Of 15 to 225.
20. $16:(?) = \frac{1}{2}$. $14:(?) = 2$.
21. $(?):5 = 4$. $(?):8 = \frac{1}{4}$.
22. Find the value of the compound ratio, $\left.\begin{array}{l}8:10\\5:6\end{array}\right\}$.

PROPORTION.

331. Oral.

23. Give three fractions having the same value as $\frac{2}{3}$.
24. Give two numbers that have the same ratio as 5 to 10.
25. Give a fraction equal to $\frac{3}{4}$.

RATIO AND PROPORTION. 191

26. Give a ratio equal to 3 : 4.

27. How does the ratio of 5 men to 10 men compare with the ratio of $5 to $10?

28. How does the ratio of 8 lb. to 4 lb. compare with the ratio of 40¢ to 20¢?

29. Name two numbers that have the same relation as 5 to 10. As 4 to 24. As 8 to 16. As $\frac{1}{2}$ to $\frac{1}{4}$.

30. What number has the same relation to 5 as 12 to 3?

31. Find a number whose ratio to 4 equals 3 : 6.

32. Give three ratios equal to $100 : $50.

33. Give any two ratios that equal each other, and express their equality.

332. An equality of ratios is a **Proportion**. Thus, 4 : 2 = 12 : 6. The ratio of 4 to 2 equals the ratio of 12 to 6.

A proportion is usually expressed with the sign (::) between the ratios; thus, 4 : 2 :: 12 : 6. This is read *4 is to 2 as 12 is to 6*.

A proportion has four terms, of which two are antecedents and two are consequents. Each term is a proportional.

333. The first and fourth terms are called **Extremes**, and the second and third terms are called **Means**.

NOTE. — In the proportion 2 : 6 :: 6 : 18, the two means are the same number, 6. The 6 is called a mean proportional.

PRINCIPLE. — The product of the extremes equals the product of the means.

Rule. — *To find an extreme, divide the product of the means by the given extreme.*

To find a mean, divide the product of the extremes by the given mean.

Supply the missing term:

34. $1 : 836 :: 25 : (\quad)$.   39. $10 \text{ yd.} : 50 \text{ yd.} :: \$20 : (\quad)$.
35. $6 : 24 :: (\quad) : 40$.   40. $15 \text{ lb.} : 60 \text{ lb.} :: (\$ \ ) : \$12$.
36. $(\quad) : 15 :: 60 : 6$.   41. $\tfrac{1}{2} \text{ da.} : (\text{ da.}) :: 12 : 6$.
37. $25 : (\quad) :: 4 : 8$.   42. $(\text{ men}) : 75 :: \$50 : \$150$.
38. $6 : 4 :: \tfrac{1}{3} : (\quad)$.   43. $\$\tfrac{3}{4} : \$3\tfrac{3}{4} :: (\quad) : 5$.

SIMPLE PROPORTION.

334. An equality of two simple ratios is a **Simple Proportion.**

It is employed in solving questions having three given terms, two of which have the same relation to each other as the third to the required term.

44. If 12 bushels of oats cost $4, what will 60 bushels cost?

SOLUTION. — There must be the same relation between the *cost* of 12 bu. and the *cost* of 60 bu. as exists between 12 bu. and 60 bu.

$12 : 60 :: \$4 : (\$ \quad)$

$\dfrac{60 \times 4}{12} = \$20$.

We place $4 for the third term. The answer will be the fourth. We must now form a ratio of 12 and 60 that shall equal the ratio of $4 to the answer. Since the third term is less than the required answer, the first must be less than the second, and we have 12 : 60 for the first ratio. The product of the means divided by the given extreme will give the other extreme, or $20. *Ans.*

By analysis, — Since 12 bu. cost $4,
    1 bu. will cost $⅓, and
    60 bu. will cost $20. *Ans.*

Rule. — *Consider the required answer as the fourth term, and place the number that is like it for the third term.*
Place the two remaining terms as follows:
If the answer is to be larger than the third term, the second must be larger than the first. If smaller, the second must be smaller than the first.
Divide the product of the means by the given extreme. Cancel when possible.

RATIO AND PROPORTION.

45. If 10 sheep cost $35, what will 23 sheep cost? What will 6 sheep cost?

46. If 5 men can do a piece of work in 9 days, how long will it take 15 men to do the same work?

SOLUTION. — Place 9 days for the third term, because it is like the required answer, thus,

:: 9 da. : (da.)

Since 5 men can do it in 9 days, 15 men can do it in less time. Therefore, since the answer is to be smaller than the third term, place 5 men for the second, and 15 men for the first. Multiplying and dividing we have 3 days, *Ans.*

47. If 14 horses eat 36 tons of hay in a certain time, how many tons will 13 horses eat in the same time?

48. If it costs $400 to lay 80 rods of street-car track, how much will it cost to lay $3\frac{1}{2}$ miles at the same rate?

49. If a pole 8 ft. high casts a shadow $4\frac{1}{2}$ ft. long, how high is a tree which casts a shadow 48 ft. long?

50. If a man walks 280 miles in 8 days, how many days ought it to take him to walk 420 miles?

51. If it costs $13.20 to supply a new arithmetic to each of a class of 24 pupils, what will be the expense of furnishing one to each of a class of 19?

52. How far can a train run in 3 hours, if it can run 160 Km. in 4 hours?

53. How many men will be required to do in 10 days what 15 men can do in 30 days?

54. What will 8 tons of coal cost, when $17\frac{1}{2}$ tons cost $78.75?

55. If a certain sum of money yields $360 interest in one year, what would the interest of the same sum be for 15 months?

56. If $800 yield $48 interest in a certain time, how large a sum will yield $216 in the same time?

57. If the interest of $3600 for a certain time is $216, what will be the interest of $800 for the same time?

58. If a garrison of 240 soldiers have a supply of food sufficient for 150 days, how long would the same food last if the garrison were increased to 600 men?

59. In the above example, how long would the food last if 80 men were sent away?

60. Find the cost of $1\frac{1}{2}\frac{2}{?}$ bushels of wheat, if $\frac{2}{7}$ bu. costs $\$\frac{6}{11}$.

61. If 120 shoemakers make 40 dozen pair of shoes in a certain time, how many shoemakers would it require to make the same number of shoes in one-half of the time?

62. If a train runs 140 miles in 4 hr. 30 min., what is the rate per hour?

63. When 5 tons 1650 lb. of coal cost $24.75, what will be the cost of 18 tons 675 lb.?

64. If a 16-foot board 9 inches wide contains 12 sq. ft., how wide must a board of the same length be to contain 20 sq. ft.?

65. It takes 26 yards of carpet 1 yard wide to cover a floor. How many yards will it take if the carpet is but 27 inches wide?

66. The ratio of Simon's pay to Matthew's is $\frac{2}{3}$. Simon earns $18 per week; what does Matthew earn?

67. 25 men can do a piece of work in 70 days; but after 30 days 15 of them refuse to work. In how many days can the rest complete the work?

RATIO AND PROPORTION.

COMPOUND PROPORTION.

335. An equality between a compound and a simple ratio is a **Compound Proportion**; thus,

$\left.\begin{array}{l}8:4\\3:10\end{array}\right\} :: 12:20$ is a compound proportion.

Find the fourth term.

SOLUTION. — First changing to a simple proportion, we have,

$\left.\begin{array}{l}3:6\\4:8\end{array}\right\} :: 3:(\quad)$

$3 \times 4 : 6 \times 8 :: 3 : (\quad)$.

Then divide the product of the means by the given extreme, using cancellation. Thus,

$$\frac{6 \times \cancel{8} \times \cancel{3}}{\cancel{3} \times \cancel{4}} = 12. \ Ans.$$
(with 2 above the 8, and the 3's and 4 cancelled)

1. If 5 men earn $72 in 8 days, how much can 10 men earn in 6 days?

SOLUTION. — Since the answer is to be in dollars, place $72 for the third term, and arrange the terms of each couplet according as the answer should be greater or less than the third term if it depended on that couplet alone.

$\left.\begin{array}{l}\text{5 men : 10 men}\\\text{8 days : 6 days}\end{array}\right\} :: \$72 : (\quad)$

Since 5 men earn $72, 10 men can earn more, so we place 10 men for the second and 5 men for the first; and since they earn $72 in 8 days, they will earn less in 6 days, so we place 6 days for the second term, and 8 days for the first. Dividing the product of the means by the extremes, we have,

$$\frac{\cancel{\$72} \times \cancel{10}^{2} \times 6}{\cancel{5} \times \cancel{8}} = \$108. \ Ans.$$

By analysis.

Since, . 5 men in 8 days earn $72;
Therefore, 1 man in 8 days will earn $\frac{72}{5}$;
1 man in 1 day will earn $\frac{9}{5}$;
10 men in 1 day will earn $\frac{90}{5}$;
10 men in 6 days will earn ($108).

Rule. — *Consider the answer as the fourth term, and place the number that is like it for the third.*

Arrange the couplets as if the answer depended on each couplet alone, as in simple proportion.

Divide the product of the means by the product of the extremes.

Cancel when possible.

2. If 4 horses eat 10 bushels of oats in 5 days, how many bushels will be required to feed 5 horses for 4 days?

3. If 10 men working 8 hours a day can do a piece of work in 12 days, how many days would it take 6 men, working 10 hours a day, to do the same amount of work?

4. If a wheelman rides 144 miles in 3 days of 6 hours each, how many miles can he ride in 5 days of 9 hours each?

5. A section of a street 33 feet long and 20 feet wide can be paved with 15840 stones, each 9 inches long and 8 inches wide. How many stones 12 inches long and 10 inches wide will it take to pave a street 12 rods long and 16 feet wide?

6. If it costs $84 to carpet a room 24 feet long and 21 feet wide with carpet 27 inches wide, how much will it cost to carpet a room 25 feet long and 18 feet wide with carpet 1 yard wide?

7. If 18 men chop 360 cords of wood in 12 days of 9 hours each, how many cords could 17 men chop in 13 days of 10 hours each?

8. If 50 men, working 10 hours a day for 11 days, can dig 25 rods of a canal 60 ft. wide, 5 ft. deep, how many rods of a canal 90 ft. wide, 7 ft. deep, can 140 men dig in 22 days of 8 hours each?

9. If 60 men can build a wall 150 ft. long, 64 ft. high, 2 ft. thick, in 8 days of 10 hours each, how many days of 8 hours each will 36 men require to build a wall 180 ft. long, 80 ft. high, $2\frac{1}{2}$ ft. thick?

10. How many men will it require to mow 48 acres in 3 days of 12 hours each, if 6 men mow 24 acres in 4 days of 9 hours?

11. If 4 lb. 6 oz. of tea cost $2\frac{3}{16}$, what will 3 lb. 11 oz. cost at same rate?

12. If sufficient flour to fill 8 bags containing 98 lb. each can be produced from 16 bushels of wheat, how many bushels will be needed to fill 14 barrels of 196 lb. each?

13. My gas bill for the month of November is $3.50 when I use 6 burners $3\frac{1}{2}$ hours each evening. How much ought it be for the month of December, when I use 4 burners for 5 hours each evening?

14. How long a piece of cloth .4 m. wide, can be made from 175 Kg. of wool, if 45 Kg. make a piece 25 m. long and .6 m. wide?

15. How many hours daily ought 30 men to labor to perform in 10 days a piece of work which is $\frac{2}{3}$ as great as a similar job which 25 men, working 12 hr. per day, accomplished in 12 days?

16. If $475 yield $171 interest in 6 years, how long will it take $960 to double itself at the same rate?

17. A bin 8 ft. long, 6 ft. wide, and $4\frac{1}{2}$ ft. deep will contain 270 bushels of wheat. How deep must another bin be built, that is 12 ft. long and 9 ft. wide, to hold 405 bushels?

18. How many days ought it to take 9 men to build a wall 350 feet long, $2\frac{1}{2}$ feet high, and 3 ft. thick, if 10 men build a wall 312 ft. long, 6 ft. high, and 2 ft. thick, in 30 days?

PARTNERSHIP.

336. Oral.

1. Charles and John share $28 in the ratio of 2 to 5. How much has each?

SOLUTION. — Charles has $2 as often as John has $5. Both have $7. Charles has $\frac{2}{7}$ and John $\frac{5}{7}$ of $7. Since they have respectively $\frac{2}{7}$ and $\frac{5}{7}$ of a *part* of $28, they must have the same fractions of the whole. Therefore,

Charles has $\frac{2}{7}$ of $28 = $8. ⎫
John has $\frac{5}{7}$ of $28 = $20. ⎬ *Ans.*

2. Divide 30 into two such parts as shall be to each other as 7 to 8.

3. A man and a boy together earn $48. The man has earned $3 to the boy's $1. What is each one's share?

4. A horse and a cow were bought for $150. The horse cost twice as much as the cow. What was the cost of each?

5. Divide 140 into four parts that shall be to each other as 2, 3, 4, and 5.

6. A father divided $7200 among his three sons in proportion to their ages, which were 10, 12, and 14 years respectively. What was the share of each?

7. A man divided $3.60 among three boys, giving to the first 5 cents as often as he gave 6 cents to the second and 7 cents to the third. How much did each boy receive?

8. Professor Adams caught 520 fish in a season, consisting of trout, black bass, and pickerel, in the proportion of 5, 4½, and 3½. How many of each kind did he catch?

337. The association of two or more persons in business is called **Partnership**.

The persons associated are **Partners**.

PARTNERSHIP.

The association is called a **Firm**, or **Company**.
All the money or property furnished by the partners constitutes the **Capital**.

338. To find each partner's share of the Gain or Loss, when their capital is employed for the same time.

1. A, B, and C formed a partnership. A contributed to the capital $800; B, $1000; and C, $1200. At the end of a year they found that there was a gain of $1500. What was each man's share of the gain?

SOLUTION. —
$800 + $1000 + $1200 = $3000, entire capital.
A's gain, $\frac{800}{3000}$, or $\frac{4}{15}$ of $1500 = $400.
B's gain, $\frac{1000}{3000}$, or $\frac{5}{15}$ of $1500 = $500.
C's gain, $\frac{1200}{3000}$, or $\frac{6}{15}$ of $1500 = $600.

Rule. — *Take for each man's share of the gain or loss such a part of the whole gain or loss as his capital is of the whole capital.*

2. Mr. Wilson and Mr. Mead entered into partnership. Mr. Wilson's capital was $3000, and Mr. Mead's $2000. They gained $1500. What was each partner's share of the gain?

3. Messrs. Jones and Smith are partners, with a capital of $3000 and $5000 respectively. After one year they find that they have gained $2000. How much of the gain should each receive?

4. Three men form a partnership at the same time. A invests $1250; B, $2000; C, $1550. They gain $1200. What is each man's share of the gain?

5. Three men hired a coach to convey them to their respective homes. A's home was 20 miles away, B's 24 miles, and C's 28 miles. They paid $24 for the coach. What ought each to pay?

6. A cargo of wheat valued at $4500 was entirely destroyed. One-third of it belonged to A, two-fifths to B, and the remainder to C. What was each one's share of the loss, there being an insurance of $3600?

7. A man fails in business to the amount of $15000, and his available means amount to only $9000. How much will two of his creditors receive, to one of whom he owes $3000, and the other $4500?

8. A and B gain in business $2500, of which A's share is $1000, and B's $1500. What part of the capital does each furnish? and what is the investment of each if their joint capital is $16000?

9. I form a partnership with two members of my class. The second member invests a certain amount, the first invests $\frac{1}{2}$ as much, while I invest as much as the other two. What share of the profit do I get?

10. Purchased a flour-mill for $42000. X's share of the mill was $\frac{5}{12}$, Y's $\frac{1}{4}$, and Z's the remainder. At the end of three years they sold the mill at a reduction of $5000, but the profits in the business during the three years were $20000. What was each man's net gain?

11. Two persons have invested in trade $800. They gain $150. The gain and stock of the first amount to $570. What is the stock and the gain of each?

When the capital of the partners is not employed for the same time.

A and B formed a partnership. A furnished $500 for 8 months, and B $600 for 10 months. They gained $360. What was each partner's gain?

SOLUTION. — A $500 × 8 mo. = $4000 for 1 mo.
B $600 × 10 mo. = 6000 for 1 mo.
$10000

PARTNERSHIP.   201

A's share = $\frac{4}{10}$ of $360 = $144.
B's share = $\frac{6}{10}$ of $360 = $216.

The use of $500 for 8 months is equivalent to the use of $4000 for 1 month; and the use of $600 for 10 months is equivalent to the use of $6000 for 1 month. Consider A's capital to be $4000 and B's $6000 = A's share of gain, $\frac{4}{10}$; B's share of gain, $\frac{6}{10}$.

2. A commenced business with $10000 capital. Four months later B put in $10500. Their profits at the end of a year were $5100. What was each man's share of the gain?

3. Three persons loaned a sum of money for which they received $1596 interest. The first had $4000 invested for 12 mo., the second $3000 for 15 mo., and the third $5000 for 8 mo. How much interest did each receive?

4. A and B were in partnership for 2 years. A at first invested $2000, and B $2800. At the end of 9 months A took out $700, and B put in $500. They lost in the two years $3720. Apportion the loss.

5. A, C, and H form partnership. A puts in $8000, C $5000, H $10000. A's capital remains in the business 8 mo., C's 9 mo., H's 12 mo. The net gain is $6900. Find each man's share of the gain.

6. Two partners entered business, agreeing to continue for 18 months. A put in $2000 at first, and 8 months later $1200 additional. B at first put in $3000, but at the end of 4 months drew out $600. On closing their account they found they had made $2808. What was each man's share of the gain?

7. On Feb. 1 Messrs. Scott and White commenced business with $3000, each furnishing $1500. On April 1 White put in $1300 more. On May 1 they took Watson into partnership with $2500. At the close of the year, how should a net gain of $2400 be apportioned?

8. A's capital was in business 6 months, B's 7 months, and C's 11 months. A's gain was $600, B's $1400, and C's $990. Their joint capital was $7800. What was each man's capital?

9. A put $600 in trade for 5 months, and B $700 for 6 months. They gained $228. What was each man's share?

10. April 1, 1895, A goes into business with a capital of $6000; July 1, he takes in B as a partner with a capital of $8000; and Oct. 1, 1896, they have gained $2900. Find the gain of each.

11. Three men, A, B, and C, hire a pasture for 6 months for $75. A puts in 10 cows at first, but at the end of 1 month takes away 4. B puts in 8, and in 3 months takes out 5, but adds 2 after 2 months more. C puts in 6, and in 4 months he puts in 8 more. What should each pay?

QUESTIONS.

339. 1. Define ratio; the terms of a ratio.

2. How is ratio found? What is direct ratio? Inverse ratio? A simple ratio? A compound ratio?

3. Tell how to find ratio when antecedent and consequent are given. To find consequent when antecedent and ratio are given. To find antecedent when consequent and ratio are given. What are the principles of ratio?

4. Define proportion. How many terms in a simple proportion? Name them.

5. Give the principles of proportion.

6. What number is placed for the third term? The second? The first? How is the fourth term then found?

7. What is a compound proportion? What number is placed as the third term? How is each couplet then arranged? How find the fourth term?

8. What is a partnership? A company?

9. Define capital stock; dividends.

10. Tell how to find each partner's share when the capital of each is invested for the same time. When the capital of each is not invested for the same time.

INVOLUTION.

340. 1. $3 \times 4 \times 2 =$ what?

2. $3 \times 3 \times 3 =$ what?

NOTE.— In Ex. 2 the factors are equal; in Ex. 1 they are unequal.

The product of equal factors is a **Power**.

3. What is the product of 4 taken 3 times as a factor?

4. What is the product of 6 taken twice as a factor?

5. What is the product of $\frac{2}{3}$ used three times as a factor?

6. What is the product of .6 used twice as a factor?

341. The process of finding powers is **Involution**.

342. A power is named according to the number of its equal factors.

The product of *two* equal factors is the **Second Power**, or **Square**, of the equal factor.

The product of *three* equal factors is the **Third Power**, or **Cube**, of the factor.

NOTE. — The second power is called a square because the area of any square figure is the product of two equal factors, length and breadth; and the third power is called a cube because the solidity of any cube is the product of three equal factors, length, breadth, and thickness.

343. A small figure at the right and above a number to show how many times it is to be used as a factor is called an **Exponent**. Thus,

$4^2 = 4 \times 4$ is 4 to the second power, or the square of 4;
$2^3 = 2 \times 2 \times 2$ is 2 to the third power, or the cube of 2;
$3^4 = 3 \times 3 \times 3 \times 3$ is 3 to the fourth power, or the fourth power of 3.

Read: 8^2, 15^8, 5^7, $(3/4)^2$, $3/4^2$, $3^2/4$, 84^{10}, 16^3.

344. Find the powers:

7. 5^3.
8. 2^4.
9. 25^2.
10. 6^5.
11. 1^4.
12. $.01^4$.
13. 2.5^3.
14. 1.1^2.
15. $.002^3$.
16. $(\tfrac{2}{3})^3$.
17. $(\tfrac{4}{9})^2$.
18. $(2\tfrac{1}{2})^3$.

EVOLUTION.

345. 1. What factor is used 3 times to produce 27?
2. What are the two equal factors of 64?
3. What is one of the three equal factors of 8?
4. 36 is the square of what number?
5. 64 is the cube of what number?
6. 144 is the second power of what?
7. 1728 is the cube of what?

346. One of the equal factors of a power is a **Root**.

One of two equal factors of a number is the **Square Root** of it.

One of the three equal factors of a number is the **Cube Root** of it.

The fourth root of a number is one of its 4 equal factors.

The square root of $16 = 4$. The cube root of $27 = 3$. The fourth root of $16 = 2$.

347. The **Radical Sign** ($\sqrt{\ }$) placed before a number indicates that its root is to be found.

The radical sign alone before a number indicates the square root; thus, $\sqrt{9} = 3$ is read, the square root of $9 = 3$.

SQUARE ROOT.

348. A small figure placed in the opening of the radical sign is called the **Index** of the root, and shows what root is to be taken; thus, $\sqrt[3]{8} = 2$ is read, the cube root of 8 is 2. Read the following:

$\sqrt{81}$, $\sqrt[3]{64}$, $\sqrt[4]{81}$, $\sqrt{144}$, $\sqrt[3]{1728}$, $\sqrt[4]{9}$, $\sqrt[3]{3.64}$.

EVOLUTION AND INVOLUTION.

349. 1. Find the square of 11. The cube of 6. The fourth power of 5.

2. Find the square root of 49. The cube root of 8. The square root of $\frac{9}{16}$.

3. $9^2 = ?$ $\sqrt[2]{9} = ?$ $8^3 = ?$ $\sqrt[3]{8} = ?$

4. Write all the squares from 1 to 100.

5. Write all the cubes from 1 to 1000.

6. Learn the second and third powers of numbers from 1 to 12.

SQUARE ROOT.

350. The square of a number is the product of that number taken twice as a factor.

Blackboard.

$1^2 = 1.$ $10^2 = 100.$ $100^2 = 10000.$
$9^2 = 81.$ $90^2 = 8100.$ $900^2 = 810000.$

From the above illustration it is seen that annexing one cipher to a number annexes two ciphers to the square of that number, as in $1^2 = 1$; $10^2 = 100$; $100^2 = 10000$.

351. A square contains twice as many figures as its root, or twice as many less one.

Squares of even tens.

Oral.

1. $20^2 = ?$ 3. $80^2 = ?$ 5. $70^2 = ?$ 7. $500^2 = ?$ 9. $600^2 = ?$
2. $50^2 = ?$ 4. $30^2 = ?$ 6. $200^2 = ?$ 8. $900^2 = ?$

352. The square of a number composed of tens and units may be found as follows:

$$24 = 20 + 4 = 2 \text{ tens} + 4 \text{ units}.$$
$$24^2 = (20 + 4) \times (20 + 4).$$

$$
\begin{array}{r}
20 + 4 = 24 \\
20 + 4 = \underline{24} \\
(20 \times 4) + 4 = \overline{96} \\
20^2 + (20 \times 4) = \underline{480} \\
20^2 + 2 \times (20 \times 4) + 4^2 = \overline{576}
\end{array}
$$

From the operation, we find that,

$$
\begin{array}{ll}
\text{The square of the tens} & 20 = 400 \\
\text{2 times the tens by the units,} & 2 \times (20 \times 4) = 160 \\
\text{The square of the units} & 4 = \underline{16} \\
& 400 + 160 + 16 = \overline{576}
\end{array}
$$

353. Principle. — The square of a number composed of tens and units is equal to the square of the tens, plus twice the product of the tens by the units, plus the square of the units.

Formula. — $\text{Tens}^2 + 2 \times \text{tens} \times \text{units} + \text{units}^2$.

Separate the following into tens and units, and find their squares: 15, 25, 74.

354. By reversing the process we may find the Square Root.

10. What is the square root of 1225?

Solution. — Separating into periods of two figures each, beginning at units, we have 12′25. Since there are two periods in the power, there must be two figures in the root, tens and units.

The greatest square of even tens contained in 1225 is 900, and its square root is 30 (3 tens).

$$
\begin{array}{lr}
\text{Tens}^2, 30^2 & = \\
2 \times \text{tens} = 2 \times 30 & = 60 \\
2 \times \text{tens} + \text{units} = 2 \times 30 + 5 & = 65
\end{array}
\quad
\begin{array}{|l}
1225\ \underline{\ 30 + 5 = 35.} \\
\underline{900} \\
325 \\
\underline{325}
\end{array}
$$

Subtracting the square of the tens, 900, the remainder consists of $2 \times (\text{tens} \times \text{units}) + \text{units}$.

SQUARE ROOT. 207

325, therefore, is composed of two factors, units being one of them, and 2 × tens + units being the other. But the greater part of this factor is 2 × tens (2 × 30 = 60). By trial we divide 325 by 60 to find the other factor (units), which is 5, if correct. Completing the factor, we have 2 × tens + units = 65, which, multiplied by the other factor, 5, gives 325, proving the correctness of the solution. Therefore the square root is 30 + 5 = 35.

355. Square root may be explained by the aid of diagrams.

The area of every square surface is the product of two equal factors, length and width.

Finding the square root of a number, therefore, is equivalent to finding the width of a square surface, its area being given.

356. The following formulas illustrate the principles which underlie the operations of square root:

    1. Length × Width = Area.
    2. Area ÷ Length = Width.
    3. Area ÷ Width = Length.

1. Find the width of a square whose area is 1296 sq. ft.

Fig. 1.

LENGTH.	AREA.	WIDTH.
	1296	
$30^2 =$	900	30 ft.
2 × 30 ft. = 60 ft.	396	6 ft.
2 × 30 ft. + 6 ft. = 66 ft.	396	36 ft.
	Ans.	

SOLUTION.

The greatest square of even tens contained in 1296 sq. ft. is 900 sq. ft. (Square *A*). Its width is 30 ft. 1296 sq. ft. − 900 sq. ft. = 396 sq. ft., the area of *b*, *c*, and *d*, considered as one rectangle (Fig. 2), whose width we desire to find. The length of this rectangle is (2 × 30 ft.) 60 ft. + the length of *c*. But we cannot know the length of c till we find its width. By trial (Formula 2), we divide the area, 396 sq. ft., by 60 ft., its approximate length. The quotient, if correct, is 6 ft., the width desired. To test the correctness: Add the 6 ft. to the trial divisor, and we have 66 ft., the entire length of a, b, and c, which (Formula 1), multiplied by its width, 6 ft., gives its area, 396 sq. ft. There is no remainder, and the work is correct. Therefore,

30 ft., the width of A, $+$ 6 ft., the width of a, b, and c, $=$ 36 ft., the width of the original square.

NOTES. — All the numbers in the middle column denote area. 1296 sq. ft. = area of the original square; 900 sq. ft. = the area of A; and 396 sq. ft. the area of b, c, and d.

· The numbers in the left-hand column denote length. 60 ft. = the approximate length (or the trial divisor) of b, c, and d; and 66 ft. the exact length, or the complete divisor.

The numbers in the right-hand column denote width. 30 ft = the width of A; and 6 ft. the width of b, c, and d; 36 ft. = the width of the original square.

In dividing, to find the width of b, c, and d, since the divisor is too small, care must be taken that the quotient figure be not too large.

SHORT METHOD.

357. Ex. 2. Find the square root of 1306.0996.

$$13'06'.09'96\ (36.14$$
$$9$$
$$66\ )\ 406$$
$$396$$
$$721\ )\ 1009$$
$$721$$
$$7224\ )\ 28896$$
$$28896$$

Rule. — *Beginning at the decimal point, separate the number into periods of two figures each, pointing whole numbers to the left and decimals to the right. Find the greatest square in the left-hand period, and write its*

SQUARE ROOT. 209

root at the right. *Subtract the square from the left-hand period, and bring down the next period for a dividend. Divide the dividend by twice the root already found, and annex the quotient to the root, and to the divisor. Multiply this complete divisor by the last root figure, and bring down the next period for a dividend, as before.*
Proceed in this manner till all the periods are exhausted.

NOTE 1. — When 0 occurs in the root, annex 0 to the trial divisor, bring down the next period, and divide as before.

NOTE 2. — If there is a remainder after all the periods are exhausted, annex decimal periods.

NOTE 3. — If, after multiplying by any root figure, the product is larger than the dividend, the root figure is too large and must be diminished. Also the last figure in the complete divisor must be diminished.

NOTE 4. — For every decimal period in the power, there must be a decimal figure in the root.

NOTE 5. — If the last decimal period does not contain three figures, supply the deficiency by annexing one or more ciphers.

Ex. 3. Find the square root of 253009.

SOLUTION.

25'30'09 (5   As 0 occurs in the   25'30'09 (503 *Ans.*
25   root, we annex 0 to the   25
10) 30   trial divisor, 10, and an- 1003) 3009
                    other period to the divi-   3009
dend, and divide as before. Thus, —

NOTE. — To find the square root of a common fraction, extract the root of each term separately. If both terms are not squares, change the fraction to a decimal, and then extract the root. The result will be the approximate root. Change mixed numbers to improper fractions.

4. What is the square root of $\frac{81}{144}$? $\frac{\sqrt{81}}{\sqrt{144}} = \frac{9}{12}$. *Ans.*

Find the square root of:

5. 8836.	14. .06432	23. $\frac{90}{100}$
6. 15876.	15. .005625	24. $\frac{20}{45}$
— 7. 370881	— 16. .913936	25. $\frac{256}{324}$
8. 46656	17. 25.6036	26. $4\frac{1}{9}$
9. 820836	18. 24.3049	— 27. $\frac{1}{3}$
10. 29.0521	19. .612089	28. $\frac{3969}{5625}$
11. 9.2416	20. 329.7643217	29. $36.45\frac{3}{4}$
12. 3180.96	21. 1684.298431	— 30. $2863\frac{17}{56}$
— 13. .007921	22. 389765268	31. $189\frac{13}{49}$

Find the square root to four decimal places:

32. .15	36. 72.5	40. 963
33. .18	37. 119	41. $13.2\frac{5}{7}$
34. 17	38. 3.67	42. $.009\frac{9}{16}$
35. 4.7	39. .222	43. $.003\frac{3}{7}$

44. What is the length of one side of a square field that has an area equal to a field 75 rd. long and 45 rd. wide?

45. How wide is a field containing 7056 square rods?

Perform the indicated operations.

NOTE. — Carry decimals to the third place.

46. $\sqrt{3.26 \times .0063}$.

— 47. $.03 \times \sqrt{\frac{5}{8} + \frac{9}{7}}$.

48. $\frac{1}{\sqrt{9}} \times \frac{\sqrt{9}}{3}$.

49. $\sqrt{\frac{1}{4} \times \frac{1}{9}}$.

50. $\sqrt{\frac{1}{4}} \times \sqrt{\frac{1}{9}}$.

51. $\sqrt{3.532 \div 6.28}$.

52. $\sqrt{4 + 6^2 + 2}$.

— 53. $\sqrt{\frac{1}{2}^5 + (\frac{3}{4})^3}$.

54. $\sqrt{625 + 1296}$.

— 55. $\sqrt{625} + \sqrt{1296}$.

RIGHT-ANGLED TRIANGLES.

358. A triangle having one right angle is a **Right-Angled Triangle**.

359. The side opposite the right angle is the **Hypothenuse**, as AB. BC is the **Perpendicular**, and AC the **Base**. In the triangle ABC, the hypothenuse is 5 inches, the perpendicular 3 inches, and the base 4 inches.

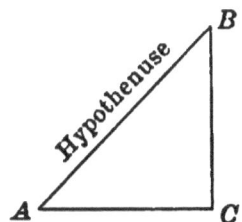
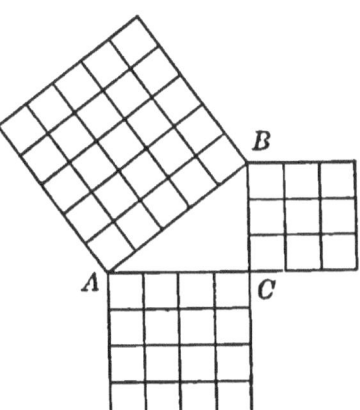

360. It will be seen that the square of the hypothenuse is 25 sq. in., which is equal to the square of the base, 16 sq. in., plus the square of the perpendicular, 9 sq. in.

PRINCIPLE. — The square of the hypothenuse equals the sum of the squares of the two shorter sides. Therefore, to find the hypothenuse, take the square root of the sum of the squares of the base and perpendicular.

$$\sqrt{\text{Base}^2 + \text{Perpendicular}^2} = \text{Hypothenuse}.$$

361. To find the base or the perpendicular, take the square root of the difference between the squares of the hypothenuse and the other side.

$$\sqrt{\text{Hypothenuse}^2 - \text{Base}^2} = \text{Perpendicular}.$$

$$\sqrt{\text{Hypothenuse}^2 - \text{Perpendicular}^2} = \text{Base}.$$

1. The base of a right-angled triangle is 32 ft., and the perpendicular 24 ft. What is the hypothenuse?

SOLUTION.— $32^2 + 24^2 = 1600$. $\sqrt{1600} = 40$ ft. *Ans.*
Or, $\sqrt{32^2 + 24^2} = 40$ ft.

2. The hypothenuse of a right-angled triangle is 40 ft., and the base 32 ft. What is the perpendicular?

$40^2 - 32^2 = 576$. $\sqrt{576} = 24$ ft. *Ans.*
Or, $\sqrt{40^2 - 32^2} = 24$ ft.

3. A 40-foot ladder placed 24 feet from a house will just reach to the top of it. How high is the house?

4. What is the length of a ladder that will reach the top of a house 40 feet high, when the foot is placed 30 feet from the house?

5. A rope 150 ft. long fastened to the top of a flag-pole reaches the ground 40 feet from the base. How high is the pole?

6. What is the hypothenuse of a right-angled triangle whose perpendicular is 36 feet, and whose base is 27 feet?

7. A square farm contains 360 acres. What is the diagonal distance between its opposite corners?

8. A telegraph pole 32 feet high casts a shadow 28 feet in length. What is the distance from the top of the pole to the end of the shadow?

9. The base of a right-angled triangle is 16 m., and the perpendicular is 12.8 m. What is the hypothenuse?

10. A boy rides his wheel due north at the rate of 15 miles an hour, and another boy starting from the same place, rides due east at the rate of 18 miles an hour. How far are they apart at the end of 5 hours?

11. What is the length of the diagonal of a room 16 ft. long and 12 ft. wide?

12. A crayon box is 6 in. long, 4 in. wide, and 4 in. high. What is the diagonal distance across the bottom, between the opposite corners?

13. A street is 32 ft. wide from curb to curb. A telegraph pole 40 ft. high stands upon one side of the street. How long must a wire be to reach from the top of the pole to the opposite side of the street at the curb?

SIMILAR SURFACES.

362. Surfaces having the same form without regard to size are **Similar Surfaces**.

NOTE. — Any two squares or any two circles of different size are **Similar Figures**. Rectangles, triangles, etc., are similar when their corresponding dimensions are proportional.

Oral.

1. What is the area of a square whose side is 2 ft.?
2. What is the area of a square whose side is 3 ft.?
3. What is the ratio of the two sides?
4. What is the ratio of the two areas?
5. Are these ratios equal? (2 ft. : 3 ft.) (4 sq. ft. : 9 sq. ft.)

SOLUTION. — From the illustration it will be seen that the areas are to each other as the *squares* of the sides; not as 2 to 3, but as 4 to 9.

PRINCIPLES. — Similar surfaces are to each other as the squares of their corresponding dimensions.

Corresponding dimensions are to each other as the square roots of their areas.

6. A circle is 4 inches in diameter; another is 8 inches in diameter. What is the ratio of their areas?

7. A circle has an area of 16 square feet; another has an area of 64 square feet. What is the ratio of their diameters?

8. The area of a rectangle 12 ft. long is 84 square feet. What is the area of a similar rectangle 6 feet long?

9. Two similar fields have areas of 12 acres and 8 acres respectively; the larger is 32 rods wide? How wide is the smaller?

10. The altitudes of two similar triangles are 20 ft. and 10 ft.; the area of the smaller is 80 square feet. What is the area of the larger?

CUBE ROOT.

363. The cube of a number is the product of that number taken three times as a factor.

Blackboard.

$1^3 = 1.$ $10^3 = 1000.$ $100^3 = 1000000.$
$9^3 = 729.$ $90^3 = 729000.$ $900^3 = 729000000.$

364. Annexing one cipher to a number, annexes three ciphers to the cube of the number, as shown in 1^3, 10^3, 100^3, etc.

365. Cubes of even tens.

1. $10^3 = ?$ 4. $40^3 = ?$ 7. $300^3 = ?$
2. $30^3 = ?$ 5. $80^3 = ?$ 8. $800^3 = ?$
3. $50^3 = ?$ 6. $200^3 = ?$ 9. $900^3 = ?$

366. The cube of a number composed of tens and units may be found as follows:

$$24 = 20 + 4 = 2 \text{ tens} + 4 \text{ units};$$
$$24^3 = (20 + 4) \times (20 + 4) \times (20 + 4).$$

$$
\begin{array}{rr}
20 + 4 = & 24 \\
20 + 4 = & 24 \\
\hline
(20 \times 4) + 4^2 = & 96 \\
20^2 + (20 \times 4) = & 480 \\
\hline
20^2 + 2 \times (20 \times 4) + 4^2 = & 576 \\
20 + 4 = & 24 \\
\hline
(20^2 \times 4) + 2 \times (20 \times 4^2) + 4^3 = & 2304 \\
20^3 + 2\,(20^2 \times 4) + (20 \times 4^2) = & 1152 \\
\hline
20^3 + 3 \times (20^2 \times 4) + 3 \times (20 \times 4^2) + 4^3 = & 13824
\end{array}
$$

From the operation we find that,

The cube of the tens	$20^3 = 8000$
3 times the square of tens by units . . .	$3\,(20^2 \times 4) = 4800$
3 times the tens by the square of the units,	$3\,(20 \times 4^2) = 960$
The cube of the units	$4^3 = 64$

$$8000 + 4800 + 960 + 64 = 13824$$

367. Principle. — The cube of a number composed of tens and units is equal to the cube of the tens plus 3 times the square of the tens by the units, plus 3 times the tens by the square of the units, plus the cube of the units.

Formula. — $\text{Tens}^3 + 3 \times \text{tens}^2 \times \text{units} + 3 \times \text{tens} \times \text{units}^2 + \text{units}^3$.

10. Separate the following into tens and units, and find their cubes: 35, 54, 63.

368. By reversing the process, we may find the cube root.

11. What is the cube root of 13824?

Solution. — Separating into periods of three figures each, beginning at units, we have 13′824. Since there are two periods in the power, there must be two figures in the root, tens and units.

The greatest cube of even tens contained in 13824 is 8000, and its cube root is 20 (2 tens).

$$\text{Tens}^3 = 20^3 =$$
$$3 \times \text{tens}^2 = 3 \times 20^2 = 1200$$
$$3 \times \text{tens} \times \text{units} = 3 \times (20 \times 4) = 240$$
$$\text{Units}^2 = 4^2 = 16$$
$$3 \times \text{tens}^2 + 3 \times \text{tens} \times \text{units} + \text{units}^2 = 1456$$
$$(3 \times \text{tens}^2 + 3 \times \text{tens} \times \text{units} + \text{units}^2) \times \text{units} = 5824$$

$$\dfrac{13'824\ \underline{|20+4}}{8000}$$
$$5824$$

Subtracting the cube of the tens, 8000, the remainder, 5824, consists of $3 \times (\text{tens}^2 \times \text{units}) + 3 \times (\text{tens} \times \text{units}^2) + \text{units}^3$. 5824, therefore, is composed of two factors, units being one of them, and $3 \times \text{tens}^2 + 3 \times \text{tens} \times \text{units} + \text{units}^2$, being the other. But the greater part of this factor is $3 \times \text{tens}^2$. By trial we divide 5824 by $3 \times \text{tens}^2$ (1200) to find the other factor (units), which is 4 if correct. Completing the divisor, we have $1200^2 + 3 \times (20 + 4) + 4^2 = 1456$, which, multiplied by the units, 4, gives the product, 5824 proving the correctness of the work. Therefore the cube root is $20 + 4 = 24$.

369. To find the cube root by the aid of blocks.

Finding the cube root of a number is equivalent to finding the thickness of a cube, its volume being given.

The following formulas illustrate the principles that underlie operations in cube root.

NOTE. — For convenience l, b, t, and v will represent length, breadth, thickness, and volume, respectively.

(1) $l \times b \times t = v$.   (2) $v \div (l \times b) = t$.   (3) $v \div (l \times t) = b$.   (4) $v \div (b \times t) = l$.

12. What is the thickness of a cube whose volume is 13824 cubic feet?

PRODUCT OF LENGTH AND BREADTH.	VOLUMES.	THICKNESS.
3×20^2 = 1200	13'824	20 ft.
$3 \times 20 \times 4$ = 240	8000	4 ft.
4^2 = 16	5824	24 ft.
1456	5824	

SOLUTION. — The greatest cube of even tens contained in 13824 cu. ft. is 8000 cu. ft. (Cube A.) Its thickness, therefore, is 20 ft. Subtracting 8000 (A) from 13824 leaves a remainder of 5824 cu. ft., which are added in solids of equal thickness

CUBE ROOT. 217

to three sides of A, as seen in Fig. 3. It now remains to find the thickness of the additions (b, c, d), (e, f, g), and h, which have a uniform thickness. As the solids b, c, d, form the greater part of the volume of the additions (5824 cu. ft.), and the length and breadth of each is 20 ft. (the length and breadth of A), by trial, using Formula 2, we find $5824 \div (3 \times 20^2) = 4$ ft., thickness of the additions, if correct. Knowing the thickness, which is also the breadth of e, f, g, h, we find the product of the length and breadth of e, f, $g = 3 \times$

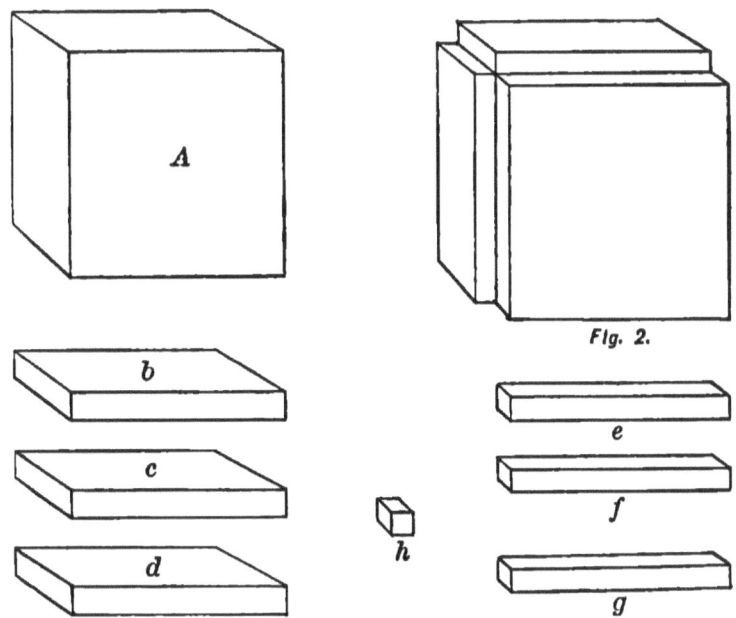

Fig. 2.

$20 \times 4 = 240$ sq. ft.; and that of $h = 4^2 = 16$ sq. ft.; both of which added to 1200 sq. ft. = the product of the length and breadth of all the additions. This product, by Formula 1, multiplied by the thickness, 4 ft., = 5824 cu. ft.; proving the correctness. Therefore,

The thickness of a cube whose volume is 13824 cu. ft. is $20 + 4$ ft. $= 24$ ft.

The numbers in the middle column (**Ex.** 12) all indicate volume:
  13824 = volume of original cube.
  8000 = volume of Cube A.
  5824 = volume of the additions (b, c, d), (e, f, g), and h.

The numbers in the left-hand column indicate product of length and breadth:

$1200 = l \times b$ of solids b, c, d.
$240 = l \times b$ of solids e, f, g.
$16 = l \times b$ of cube h.

The numbers in the right-hand column indicate thickness:

20 ft. = thickness of A.
4 ft. = thickness of all the additions.
24 ft. = thickness of original cube.

370. Short method.

Rule for finding the cube root:

Beginning at the decimal point, separate the number into periods of three figures each; thus: 16′581′.375.

Find the greatest cube in the left-hand period, and write its root at the right. Subtract the cube from the left-hand period, and bring down the next period for a dividend; thus,

```
16′581′.375 | 2
         8
      ─────
      8581
```

To find the trial divisor, square the root already found with a cipher annexed, and multiply by 3; thus,

```
                  16′581′.375 | 2
                           8       20
Trial divisor, 1200 / 8581        20
                                ─────
                                 400
                                   3
                                ─────
                                1200
```

To find the trial figure, find how many times the trial divisor is contained in the dividend; thus,

```
                  16′581′.375 | 25
                           8       20
Trial divisor, 1200 / 8581        20
                                ─────
                                 400
                                   3
                                ─────
                                1200
```

CUBE ROOT.   219

To find the correction, multiply the former root by 3, annex the trial figure, and multiply by the trial figure; thus,

```
                        16′581′.375 | 25.5
                             8        2
              1200       8581         3
               325                   65
Complete divisor, 1525   7625         5
            187500       956375     325    Continue thus, until
              3775                         all the periods are ex-
            191275       956375            hausted.
```

NOTE 1. — When there is a remainder after all the periods are exhausted, annex decimal periods, and continue the process as far as desired. The result will be the approximate root.

NOTE 2. — When a cipher occurs in the root, we annex two ciphers to the trial divisor, and bring down the next period.

NOTE 3. — The right-hand decimal period must have three places.

13. What is the cube root of 8.414975304 ?

OPERATION.

```
             8.414′975′304 | 2.034
                 8
   120000      414975              Since 0 occurs in the root, an-
     1809                          nex 00 to the trial divisor, mak-
   121809      365427              ing 120000; bring down the next
  6362700      49548304            period.
    24376
 12387076      49548304
```

NOTE. — To find the cube root of a common fraction, extract the root of each term separately. If both terms are not cubes, reduce to a decimal and then extract the root. The result will be the approximate root.

Find the cube root of
14. 42875.
15. 884736.
16. 4492125.
17. 77854483.

18. 8.615125.
19. 17.373979.
20. 450827.
21. 1879.080904.
22. 32.890033664.
23. 10077696.

24. What is the cube root of $\frac{226981}{658503}$? $\frac{7}{8}$? $1\frac{31}{8\frac{1}{3}}$? $39\frac{8}{27}$? $\frac{1}{64}$?

Extract the cube root to the third decimal place:

25. 14.323.
26. 31982.4.
27. .06324.
28. .0015.
29. 3.
30. 7.

31. What is the width of a cube whose solidity is 91125 cubic inches?

32. A cubical cistern holds 50 barrels of water. How deep is it?

33. What is the entire surface of a cube whose side is 9 ft.?

34. $\sqrt[3]{.006 \times 32.5} = ?$

SIMILAR SOLIDS.

371. Solids having the same form without regard to size are **Similar Solids**. Any two cubes or any two spheres are similar solids. Solids are similar when their corresponding dimensions are proportional.

PRINCIPLES. — Similar solids are to each other as the cubes of their corresponding dimensions.

The corresponding dimensions of similar solids are to each other as the cube roots of their volumes.

1. A globe is 3 inches in diameter, and another 6 inches in diameter. What is the ratio of their volumes?

EXPLANATION. — They are to each other as 3^3 to $6^3 = 27 : 216$.

2. There are 64 cubic inches in a 4-inch cube. How many in an 8-inch cube?

3. Two similar solids contain 386 and 284 cubic inches respectively. If the larger is 11 inches thick, how thick is the smaller?

4. If a man 6 ft. 2 in. tall weighs 215 pounds, what should be the weight of a man 5 ft. 10 in. tall of the same proportions?

5. The width of a bin is 4 ft. 6 in. How wide must a similar bin be to hold 4 times as much?

6. If an orange $2\frac{1}{2}$ inches in diameter costs 5 cents, what should an orange $3\frac{1}{2}$ inches in diameter cost?

QUESTIONS.

372. 1. What is involution? A power of a number? The first power? The second power? The third power? What are the second and third powers called? What is the exponent of a power?

2. What is evolution? A root? The square root of a number? The cube root of a number? The fourth root of a number? How is a root indicated? The square root? The fourth root?

3. Tell how to find the side of a square when the area is given?

4. Tell how to find the edge of a cube when its volume is given?

5. What kind of measure is a cube?

6. A cube contains how many times as many figures as its root?
What is shown when the number is separated into periods of three figures each?

7. What is the cube root of a number? Two answers.

8. Cube the numbers from 1 to 10.

9. What is the first root figure? What kind of measure is it?

10. How is the trial divisor found? What is the trial divisor?

11. What kind of measure is it? Why is it a trial divisor?

12. How is the correction found?

13. What kind of measure is the correction?

14. What is the complete divisor? What kind of measure is it?

15. What is a right-angled triangle?

16. What principles are true of all right-angled triangles?

17. Tell how to find hypothenuse, base, perpendicular.

18. What are similar figures? What principles are true of them?

19. What are similar solids? What principles are true of them?

REVIEW.

373. Oral.

1. What is the cost of 20 pounds of sugar at $6\frac{2}{3}$ cents a pound?

2. A man owning $\frac{2}{3}$ of a farm sold $\frac{1}{2}$ of his share. What part does he still own?

3. A can do a piece of work in 2 hours, and B in 3 hours. In what time can both do it, working together?

4. Two men receive $60 for painting a house. One worked for $2 a day, and the other $3 a day. How much money should each receive?

5. What is the interest of $500 for $2\frac{1}{2}$ years at 6%?

6. What is the cost of 64 straw hats at $1 each? At $.50? $.25? At $.12½? At $1.25? At $2.50?

7. If 4 oranges cost 12 cents, what will 7 oranges cost?

8. If ¾ of a yd. of silk costs $1½, what will 1½ yards cost?

9. If a man 6 feet tall casts a shadow 8 feet long, how long a shadow will a boy 4½ feet tall cast?

10. If ⅖ of my money is silver and the rest bills, and I have $180, how much of each kind have I?

11. If ⅔ of a cord of wood costs $1.50, what will a cord cost? 5 cords?

12. A boy buys papers at the rate of 3 for 2 cents, and sells them at the rate of 2 for 5 cents. How much does he make on 30 papers?

13. What is the value of 8 bushels of wheat, if 6 bushels cost $4.50?

14. What is the cost of 2 lb. 8 oz. of butter at 16 cents a pound?

15. What is the difference between 5 square feet and 5 feet square?

16. When it is noon in Syracuse, what time is it 7½° east of Syracuse?

17. Two places are 37½ degrees apart. When it is 5 P.M. at the eastern place, what is the time of the western?

18. When it is noon in Syracuse, what is the time 60° west of Syracuse?

19. What is the standard time of Denver when it is noon in Boston?

20. 58 is ⅔ of what number?

21. A boy sold a knife for 60 cents, which was ¾ of its cost. What did it cost?

22. The sum of two numbers is 32; their difference is 10. What are the numbers?

NOTE. — The half-sum + the half-difference = the greater. The half-sum — the half-difference = the less.

23. At a village election there were 1200 votes cast for two candidates; the successful candidate had a majority of 200 votes. How many votes were cast for each?

24. The sum of two numbers is 68; their difference is 26. What are the numbers?

25. What is $33\frac{1}{3}\%$ of $900? $66\frac{2}{3}\%$ of $1200? $12\frac{1}{2}\%$ of $96? 25% of $600?

26. A merchant, by selling goods at $80, lost 20%. What was the cost?

27. A farmer had a flock of sheep, and purchased 25% more; he then had 250 sheep. How many had he at first?

28. A lad had 45 marbles, and lost $33\frac{1}{3}\%$ of them. How many had he left?

29. What is an agent's commission for buying 96 head of cattle at $33\frac{1}{3}$ a head, at $6\frac{2}{3}\%$?

30. $75 \times 66\frac{2}{3} - 26 \times 12\frac{1}{2} = ?$

31. How much is 500% of $12?

32. A druggist expended $20 in opium, which he sold at a profit of 300%. What did he sell it for?

33. $18 is 600% of what?

34. What is the difference between .6% of $50 and $\frac{1}{2}\%$ of $70?

35. What per cent of a number is $\frac{1}{3}$ of it? $\frac{1}{2}$ of it? $\frac{1}{16}$ of it? $\frac{2}{3}$ of it? $\frac{3}{4}$ of it? $\frac{5}{8}$ of it? 16 is $\frac{1}{2}\%$ of what?

36. A lot containing 48 square rods is 3 times as long as it is wide. What are its dimensions?

EXPLANATION. — As the length is three times the breadth, we divide the area by 3; the result will be the area of each of 3 equal squares, the square root of which will be the width, which multiplied by 3 will give the length. $\sqrt{\frac{48}{3}} = 4$ rd., the width.

37. A and B had the same income. A saved $\frac{1}{4}$ of his and B $\frac{1}{8}$. A had $1600 at the end of 8 years; how much had B?

38. Which is greater, the square root of $\frac{1}{64}$, or the cube of $\frac{1}{2}$?

39. A two-inch pipe can discharge the contents of a cask in 8 hours. How long will it take a four-inch pipe?

40. How many rods of fence necessary to fence a square lot containing 144 sq. rd.?

41. A lot containing 144 sq. rd. is four times as long as it is wide. How many rods of fence does it require? (Compare with result in Ex. 40.)

42. How many inches in a hektometer?

43. How many milliliters in 4 dekaliters?

44. How many ares in 5 Hektares?

45. John and George divide 150 marbles in proportion to their ages. John is 7, and George is 8. How many marbles do each receive?

46. If a boy can ride a bicycle at the rate of 18 miles an hour, how long will it take him to ride twice around a section of land?

47. What is the interest of $600 at 8% for 3 months? for 3 years?

48. If I owe a debt of $60, and pay $40 two months before it is due, how long after it is due should the remainder be allowed to run?

49. At what time between 3 and 4 o'clock are the hour and the minute hand of a watch together?

EXPLANATION. — Both hands are together at 12 o'clock, and before it is 12 o'clock again they will have been together 11 times. They will be together between 1 and 2 in $\frac{1}{11}$ of 12 hours, and between 3 and 4 in $\frac{3}{11}$ of 12 hours.

50. If I buy 8% stock so that it pays me 6% on my investment, what per cent do I receive?

Written.

51. Frost injured 72 peach trees on M's farm, which number was 9% of all the trees he had. How many did he have in all?

52. At 2% an agent received $125.50 commission on the sale of some real estate. What was it sold for?

53.

$150.   Amsterdam, N.Y., Jan. 1, 1896.

Three months after date, I promise to pay ———— Storrie & Dunlap ———— or order, One hundred fifty Dollars, with interest. Value received.

J. W. Kimball

Find the proceeds of the above note, discounted at the Farmer's National Bank, Amsterdam, N.Y., Feb. 16, 1896.

54. A gentleman insured his house for $1800, which was ⅔ of its value, at 1¼%. In case of total destruction by fire, what is the entire loss to the owner?

55. A bill of goods amounting to $287.60 is sold with discounts of 10% and 5% for cash. How much cash will pay it?

56. If a piano that cost $360 is to be sold at a profit of $16\frac{2}{3}\%$, what price must be asked that $12\frac{1}{2}\%$ may be abated from the asking price?

57. I sold two articles for $1.50 each, thereby realizing a profit of 25% on one and a loss of 25% on the other. Did I gain or lose on both transactions?

58. A bought a carriage at 20% and 10% from list price, and sold it at 10% and 5% from list price. What per cent profit did he make?

59. A grocer bought a cask of molasses containing 40 gal. for 38 cents per gallon. Seven gallons having leaked out, for how much per gallon must he sell the remainder in order to gain $12\frac{1}{2}\%$ on the investment?

60. Suppose a grocer bought a 42-gallon cask of vinegar at 12¢ per gallon, and put 12 gallons of water with it, and sold it for the same price. What would be his rate per cent gain?

61. A meter stick is what per cent longer than a yard stick?

62. Buffalo is the largest flour depot in the world. It received by lakes and rail in 1895, 8,971,740 bbls. of flour. If the N. Y. C. & H. R.R. shipped 18.9%, the N. Y., L. E., & W. R.R. 12.15%, the Pennsylvania R.R. 8.33%, the West Shore R.R. 10.97%; the Lehigh Valley R.R. 8.12%, the other roads 6.5%, and the remainder by water, what per cent was shipped by water? and how many barrels?

63. What will be the cost of 6 loads of wood, each containing 1 C. 6 cd. ft. 10 cu. ft., at $2.50 a cord?

64. How many yards of carpet 2 ft. wide will be required for a room 12 ft. by 15 ft. 6 in., if the strips run lengthwise, and there is a waste of $\frac{1}{8}$ of a yard in each strip in matching?

65. The width of a building is 36 ft., and the ridge of the roof is 10 ft. higher than the eaves. How many square feet of boards will it take to cover one of the gable ends?

66. With how long a rope must a goat be fastened to a stake that it may feed on four square rods of land?

67. A room 24 feet long and 15 feet wide is to be carpeted with carpet $\frac{7}{8}$ yd. wide. How many yards will be required if a waste of $\frac{1}{8}$ of a yard is made on each strip in matching, the strips to run crosswise?

68. Oswego, N.Y., is in latitude 43° 28′ N. How many degrees is it from the North Pole? From the South Pole?

69. How many gallons in $32\frac{1}{2}$ hektoliters of wine?

70. If it takes 2 lb. 7 oz. 4 pwt. of silver to make 12 spoons, what amount will be required for one spoon?

71. If it is one-half of a mile from your home to the school building, how many steps of 1 ft. 6 in. each will you take in reaching it?

72. What decimal part of a week is 4 da. 3 hr. 36 min.?

73. What part of 2 reams are 10 quires, 20 sheets?

74. How many times is $132 \times 75 \times 42 \times 104$ contained in $26 \times 22 \times 150 \times 168$?

75. Bought six loads of oats, each containing 32 bags, each bag containing 2 bushels, worth $.56 a bushel, and gave in return 8 boxes of tea, each containing 24 pounds. What was the tea worth a pound?

76. $\dfrac{4 \times 7 \times 32 \times 15 \times 88}{16 \times 56 \times 5 \times 4 \times 6} = ?$

77. If $\frac{3}{4}$ of a box of oranges cost $4.50, what part of a box can be bought for $5.25?

78. Simplify the following complex fraction:

$$\frac{\frac{4}{5} \times \frac{3}{8}}{\frac{1}{3} + \frac{2}{7}} + \frac{\frac{5}{8} + \frac{3}{7}}{\frac{4}{9} \times \frac{1}{2}}.$$

79. ⅜ of 63 is $\frac{7}{12}$ of what number?

80. A gentleman invested $215380 in a knitting-mill, which was ⅔ of the value of the plant. What was the value of ⅘ of the plant?

81. A and B, being 150 miles apart, travel toward each other. They start at the same time, and meet at the end of eight hours, when they discover that A has travelled 1⅛ miles each hour more than B. How many miles has each man travelled?

82. For how long a time must $4560 be placed on interest at 6% to gain $353.40?

83. A man borrowed $250 March 3, 1896, and paid the note Sept. 21, 1896, with 5% interest. What was the amount of the note?

84. A merchant borrowed $165 at 6%, and when he paid the debt it amounted to $168.96. How long did he have the use of the money?

85. The interest on a certain sum is $27.40, the time 2 years, 3 months, 12 days, and the rate 6%. What is the principal?

86. A note for $250 was given Sept. 5, 1895; a payment of $75 was made April 25, 1896. How much will settle the note Oct. 3, 1897?

87. A man bought a farm for $4000, April 1, 1889. He gave a mortgage at 5% for $3000, and paid as follows: Jan. 1, 1890, $700; Oct. 1, 1890, $1000; April 1, 1891, $850; and the balance of the mortgage April 1, 1892. How much was due at settlement?

88. What sum of money must I loan at 6 per cent interest, that it may bring me in a quarterly income of $300?

89. Compute the interest on $3450 for 2 yr. 6 mo. 20 da. at 5%.

90. William Johnson holds a note for $1250 against James W. Way, dated Jan. 10, 1893, payable on demand. This note bears the following indorsements: March 10, 1893, $200: May 10, 1893, $300: July 10, 1893, $50; Oct. 10, 1893, $400. What is due Dec. 10, 1893, interest at 5%?

91. Find the simple interest of $382.94, one half to be paid in 5 yr. 5 mo. 20 days at 3%, the other half to be paid in 5 yr. 5 mo. 20 days at 5%.

92. A man borrows $2000 which belongs to a minor who is 18 yr. 2 mo. 10 days old, and he is to keep it until the owner is 21 years of age. What will then be due, money being worth 6%?

93. Bought a house for $6000, and gave a mortgage for $4000, dated Jan. 1, 1892, interest at 6%. Made the following payments: July 1, 1892, $520; Jan. 1, 1893, $708; Jan. 1, 1894, $680; July 1, 1895, $725. How much was due Jan. 1, 1896?

94. A man owes me $463.50, payable in 6 months without interest. What sum can I afford to take now for the debt, money being worth 6%?

95. A man bought goods amounting to $2100 on 6 mo. credit, but was offered a discount of 3% cash payment. If money was worth ½% a month, what is the difference?

96. Which is the more profitable, to buy goods worth $500 at 90 days, 3% off for cash, or put the amount at interest at 7%, and let the bill run to maturity?

97. Face of a debt, $1256.25. Date, July 1, 1886. Time, 1 yr 6 mo. Rate, 6%. What is the present worth?

98. Had a note of $2500 discounted at a Rochester bank for 2 months. What were the proceeds, rate of discount being 7%?

99. Find the difference between the true discount and the bank discount of a debt of $550, due in 4 months without interest.

100. April 1, A gave B a 3-mo. note for $300, which B had discounted at a bank May 1. What did B receive? and what amount could the bank collect on July 1, discount at 6%, no grace?

101. Bought an invoice of goods amounting to $1360.58. How much will I make by discounting my note at the bank for 90 days at 6%, and paying cash for goods at 5% off?

102. A New York note of $2000, bearing date May 24, 1895, and payable in 60 days, was discounted at 6%. The discount was $15. When was it discounted?

103. Sweet and Johonnot sold 20 bicycles to a dealer, taking his note at 60 days, which they discounted immediately at the Merchant's Bank at 6% with grace, receiving $1485. What was the price of each bicycle?

104. On the first day of January, 1890, a man gave three notes, the first for $500 payable in 30 days; the second for $400 payable in 60 days; and the third for $600 payable in 90 days. What was the average term of credit, and what the equated time of payment?

105. I wish to use $560.88 immediately. For what sum must I draw a bank note, due in 96 days at 6%, that I may receive the required amount?

106. How many $500 U. S. bonds can be bought for $6630 at $10\frac{1}{2}$% premium?

107. A guardian invests $1000 at simple interest at 3%, $1000 in 4% bonds at $112\frac{1}{2}$, and $1000 in 5% bonds at 125. The bonds are to run 10 years, and be redeemed at par. Compare the three investments at the end of the ten

108. The city of Buffalo pays $12425.72 for rented school buildings. On what amount of $3\frac{1}{2}\%$ bonds would this pay the interest?

109. If I buy bank stock at 20% discount, and sell it at 10% premium, what per cent do I gain?

110. What is the rate of income on a 4% stock bought at $62\frac{1}{2}$.

111. I have $5000 to invest, and can buy 5% stock at 110, or 6% stock at 125. Which will be the better investment? and how much annually?

112. A gentleman owned a house which he rented for $375 above all expenses. He sold the house for $5000, and invested the money in a 5% stock at 80. Did he gain or lose by the transaction? and how much per year?

113. The ratio of A's weight to that of B is $\frac{2}{3}$. B weighs 120 lb. 8 oz. What does A weigh?

114. If John is 6 years old and Henry 15, what is the ratio of John's age to that of Henry? What will it be when each is 5 years older?

115. If 4 horses eat 4 bu. of oats in 2 days, how many horses will eat 48 bushels in 12 days? (Solve by analysis.)

116. If the antecedent is $\frac{2}{3}$ of $\frac{9}{16} \times \frac{8}{45}$, and the ratio is $\frac{2}{3}$ of $\frac{9}{24}$, what is the consequent?

117. How wide can 20 men, working 8 hours a day for 8 days, make a ditch which is 75 rods long and 10 ft. deep, if 25 men, working 10 hours a day for 7 days, can dig a ditch 80 rd. long, 8 ft. deep, and 2 ft. wide?

118. If a baker's loaf weighs 10 ounces when wheat is 60 cents a bushel, what should it weigh when wheat is 70 cents a bushel?

119. If a train moves at the rate of 30 miles in 48 minutes, in what time will it run 450 miles?

REVIEW. 233

120. One side of a shed is 8 ft. high, the opposite side 13 ft. 6 in. What is the ratio between the sides?

121. If it costs $30 to lay a cement sidewalk 4 ft. wide and 16 ft. long, how much will it cost to lay the same kind of walk 7 ft. wide and $96\frac{1}{2}$ ft. long at the same rate?

122. Write and solve a problem in proportion, using the following numbers: 8 men, 9 lb. of beef, 1 da.; and 2 da., 12 lb. of beef.

123. If $\frac{3}{16}$ of a yard of cloth cost $$\frac{1}{2}$, what will $4\frac{1}{2}$ yd. cost?

124. A, B, and C, engaged in trade. A put in $400, B $250, C $600; they gain $300. Find each man's share of the gain.

125. A merchant failing in trade has debts amounting to $34560; his assets are $30240. What can he pay on the dollar? and how much will a creditor receive to whom he owes $3840?

126. A man willed his property, which was valued at $6000, to his four children in the following proportion, giving to each one $\frac{1}{2}$, $\frac{1}{4}$, $\frac{1}{3}$, and $\frac{1}{8}$ respectively. How much did each one receive?

127. Three families rent a cottage for the summer. The first family occupies it for 6 weeks, the second for 2, and the third for 3 weeks. The rent for the entire season of 11 weeks, is $440. How much should each family pay?

128. Scrantom, Morris, and Jackson were associated in business for a period of 1 yr. 6 mo. Scrantom furnished $5000, Morris $3000, and Jackson $2000 of the original capital. When the partnership terminated, they divided $4000, the profits arising from the same. How much more did each make than he would have realized had his money been invested in a 6% mortgage?

129. Divide $60 among three boys so that one shall have $\frac{1}{2}$ as much as the other two, whose shares are as 3 to 7.

130. What is the distance between the diagonally opposite corners of a lot whose area is 16 sq. ft.?

131. My dining-room is 16 ft. long, 14 ft. wide, 10 ft. high. Find diagonals of the shorter sides, of the longer sides, and of the room.

132. What is the length of one side of a cube, equal in volume to a solid that is 49 ft. long, 27 ft. wide, and 7 ft. high?

133. A ladder 25 ft. long, the bottom of which is 5 ft. from a building, reaches the base of a window. How many feet from the base of the window to the ground?

134. At 40 cents a rod for fencing, which will cost the more, to enclose a square field containing 10 A., or a field of the same area whose length is twice its width?

135. Find the cube root of 41781.923.

136. Find the square root of 41781923.

137. A cubical cistern contains 30 hhd. of water. How deep is it?

138. The volume of a rectangular prism is 200 cu. ft., and its height is 8 ft. Find its surface contents, if its two other dimensions are equal.

139. The area of a right-angled triangle is 289 sq. ft., its base is $\frac{1}{2}$ of its altitude. What is the length of its altitude?

140. If a railroad company pays 19¢ per sq. yd. for excavating, and 37½¢ per sq. yd. for drawing away the earth, what will it cost the company to remove a mound equal in volume to a cube whose side is 81 feet?

141. Forty feet directly east from a column that is 75 ft. high, I measure due north 30 ft., and find that I am in line with a stake and the column. If the stake is 25 ft. distant from my position, and 10 ft. high, what is the distance from the top of the stake to the top of the column?

142. How many rods of fence will enclose a rectangular field containing 20 acres, if the field is twice as long as it is wide? and how much will it cost at $2.45 per rod?

143. If a locomotive runs at the rate of 55 miles in 40 minutes, and its drive-wheels are 18 ft. in circumference, how many revolutions will the drive-wheel make in one hour?

144. A insured his stock for $1200. He paid a premium of $24. What was the rate of insurance?

145. Grant and Dunn bought a bill of glass amounting to $853.68, upon which they received a discount of 60%, 25%, 15%, and 2% off for cash. What was the net amount of bill?

146. My agent in Chicago sold goods to the amount of $8640. He also purchased 6800 bu. of wheat at $1.10 a bushel, paid for expenses $10.40, and received a commission of 2 ct. on every dollar. How much will he remit to me after paying all expenses?

147. A merchant buys calico at $5\frac{1}{8}$ ct. per yard, and sells at 6. What is his rate per cent of gain?

148. What is the rate of insurance when a $1000 policy for 3 years costs $7.50?

149. A man lost $13.45 on some flour by selling it at a loss of $14\frac{2}{7}\%$. What was the flour worth?

150. A farmer buys 4 tons of hay at $20 per ton, and 4 bbl. of flour at $5 per barrel. What is the cash value of the bill, if he is allowed a discount of 15%, and 5% deduction for cash?

TEST QUESTIONS.

374. Arranged, by permission, from examinations given in various cities.

1. Define addition, sum, sign of equality, subtraction, remainder, subtrahend, minuend, parenthesis, multiplication, factors, multiplicand.

2. Multiply 7258 by 395, and write each partial product in words.

3. Subtract 8969 from 9782, and prove the work.

4. Prove by an illustration that multiplication resembles addition.

5. Solve: $73.46 − ($.94 + $3.02) + $47 × 35.

6. Write in figures, XLVII.

7. Write in figures, six hundred eight thousand seventy-two.

8. Multiply 6504 by 657.

9. $\frac{7}{8}$ of 585 × 5 = ?

10. Divide 45897 by 490, and prove that your work is correct.

375. 1. Copy and find the sum: $23.17, $6043.05, $0.42, $208.97, $5486.04.

2. How many yards of linen, at 28 cents a yard, must be given for 35 bushels of potatoes, at 56 cents per bushel?

3. A man paid $13,465 for a house and some land. The house alone was worth $8,978. What was the value of the land?

4. Write this number in words, 3,782,013.

5. Write in words, XCV.; 76508.904.

TEST QUESTIONS. 237

6. How many bushels of potatoes at 50 cents a bushel will pay the entire cost of a hat at $7.50, a dress at $24, a cloak at $16.25, and gloves at $1.75?

7. 6460000 × 3000 ÷ 25000000 = ? (Use shortest way.)

8. If the dividend is 1761184 and the quotient 4684, what is the divisor?

9. What is the smallest number that will exactly contain 16, 20, 24, or 30?

10. Define dividend, remainder, product, the prime factors of a number. How do you prove division?

376. 1. Define multiplier; concrete number.

2. How can you prove an example in subtraction?

3. A merchant bought 375 bbl. of apples at $.95 a bbl.; 43 bbl. rotted; if he sells the rest at $1.10 per bbl., how much does he gain or lose on all?

4. My salary is $2350 a year, and I spend $4 a day; how much will I save in six years?

5. If 23 men own 475 bbl. of apples each, and 4 of them divide theirs equally among the rest, how many will each have then?

6. Find the prime factors of 1155.

7. If my salary is $1400 per year, and my expenses $90 per month, how long will it take me to save $4160?

8. What is the smallest quantity of grain that will fill an exact number of bins, whether they hold 312, 260, or 390 bushels?

9. What are like numbers? Give three.

10. From Albany to West Troy is 5 miles, from West Troy to Cohoes 2 miles, and from Cohoes to Saratoga is 30 miles. How far is it from Albany to Saratoga?

377. 1. Write in words 23456789.

2. Find the greatest common divisor of 75, 25, and 500, and their least common multiple.

3. If 7 tons of hay cost $105, what will be the cost of 289 tons?

4. Write the number which is composed of 3 units of the eighth order, 6 of the fifth, 2 of the third, and 9 of the second.

5. Find the contents of the smallest measure that may be filled by using either a 4-quart, a 5-quart, or a 6-quart measure.

6. Find the prime factors of 1452.

7. Solve by cancellation: A man receives $21 for 15 days' work of 7 hours each. How much should he receive for 19 days' work of 5 hours each?

8. The product of three numbers is 105840; one of the numbers is 42, the other 35. What is the third number?

9. How many pounds of butter at 20¢ a pound are worth as much as 1600 bushels of wheat at 75¢ a bushel?

10. What is the greatest common divisor of two or more numbers?

378. 1. Two persons start from the same point and travel in opposite directions; one at the rate of 25 miles a day, and the other at the rate of 32 miles a day. How far apart will they be in 8 days?

2. What is the product of 20202 × 10101?

3. What number multiplied by 1728 will produce 1705536?

4. A man has $8250; how much must he add to this to be able to pay for a farm worth $10000?

5. (6070 − 1200) + (4680 ÷ 15) = ?

6. Bought 144 acres of land at $41.25 an acre, and sold the whole for $7000. Did I gain or lose? and how much?

7. If 3 oranges are worth $\frac{2}{3}$ of a melon, what part of the melon is 1 orange worth?

8. Austin having 30 marbles, gave $\frac{1}{5}$ of them to one companion and $\frac{1}{6}$ of them to another. How many had he left?

9. How many eggs in 12½ dozen?

10. Write the present year in Roman numerals.

379. 1. George gave a beggar 9 cents, which was $\frac{1}{5}$ of all the money he had. How much money had he?

2. Mary is 14 years old, and her sister is $\frac{4}{7}$ as old. How old is her sister?

3. What is a mixed number?

4. How many ninths in 5⅜?

5. At $\frac{2}{3}$ of a dollar a pound, what will 8 pounds of butter cost?

6. What will $\frac{3}{4}$ of a pound of coffee cost at 28 cents a pound?

7. If a man earns $15 a week and spends $\frac{2}{3}$ of it, how much does he save?

8. What do you understand by $\frac{7}{8}$ of anything?

9. Change 142⅜ to an improper fraction.

10. A boy having 20 quarts of blueberries, sold $\frac{2}{3}$ of them for $$\frac{8}{20}$. What was the price for a quart?

380. 1. If I put ½ of my money in one bank, ¼ in another, ⅕ in another, and have $4,200 besides, how much have I?

2. A can mow a field in 10 days, B in 8 days, and C in 5 days. When working together, how many days will they need?

3. If 6 is added to both terms of the fraction $\frac{1}{3}$, how much is the fraction increased or diminished?

4. The divisor is 46, the quotient 605, and the remainder 23. What is the dividend?

5. From the sum of $\frac{3}{4}$ and $\frac{7}{8}$ take the sum of $\frac{4}{12}$ and $\frac{2}{3}$.

6. How many cords of pine wood at $3.25 a cord must be given for 12 yards of broadcloth at $2.10 a yard? Work and analyze.

7. Find the prime factors of 13860.

8. $\dfrac{13 \times 16 \times 42 \times 51}{6 \times 17 \times 48 \times 91} = ?$

9. Find the sum of the prime numbers under 20.

10. Reduce $\frac{1728}{1872}$ to lowest terms.

381. 1. Write in Roman notation 1894.

2. Write the prime numbers from 1 to 18 inclusive.

3. If 3 boxes of oranges cost 5\frac{2}{3}$, how many boxes can be bought for $17?

4. A farmer sold 64 sheep, and had $\frac{4}{9}$ of his flock left. How many had he left?

5. $\dfrac{\frac{7}{8} \text{ of } 3\frac{2}{3}}{\frac{3}{8} \text{ of } 2\frac{1}{4}} = ?$

6. How many barrels of flour at $6 a barrel must be given for 3 pieces of linen, each containing 36 yds., at 25 ct. a yard?

7. A farmer sold at market 15 sheep at 2\frac{5}{8}$ each, and bought 7 yards of cloth at 1\frac{3}{8}$ per yard. How much money did he take home?

8. From the sum of $5\frac{1}{2}$, $9\frac{3}{4}$, $11\frac{1}{8}$, take the difference between 32 and $13\frac{3}{8}$.

9. Reduce to their least common denominator $1\frac{6}{4}$, $2\frac{7}{6}$, $2\frac{1}{2}$, $3\frac{3}{4}$.

10. Write a receipt for $20 paid you by Mr. John Dixon.

382. 1. Add $49.50, $43.62½, $75.05, $64.75, $35.09, $6.03½, $42, $73.98, $105.60.

2. How many sheep at $5 each must be given for 15 horses at $150 each?

3. What is the sum of $\dfrac{4\frac{1}{2}}{6} + \dfrac{6\frac{1}{4}}{3} + \dfrac{\frac{3}{4}}{4}$?

4. From $\frac{2}{3}$ of $\frac{1}{4}$ of 3 take $\frac{3}{4}$ of $1\frac{1}{2}$.

5. At $39¾ apiece, how many cows can be bought for $2504¼?

6. How many times is $\frac{3}{10}$ of $\frac{5}{8}$ of 6½ contained in $\frac{5}{8}$ of $54 \times \frac{2}{3} \div \frac{1}{8}$?

7. Define multiple and greatest common divisor.

8. Give and define proper fraction; mixed number.

9. If a man spends ⅕ of his money for a house, ⅖ for a farm, and has $3400 in cash left, what is the amount of his wealth?

10. A grocer bought 3 barrels of apples of different qualities at $2.75, $3.12, and $3.25 a barrel. What was the average cost?

383. 1. What is reduction of fractions?

2. Reduce $1\frac{1}{2}$ to 156ths.

3. Express $\frac{288}{1728}$ in its simplest form.

4. Change to fractions having the least common denominator, $\frac{7}{8}$, $\frac{9}{10}$, and $\frac{3}{18}$.

5. If a merchant buys tea at $\$\frac{2}{3}$ a pound, and sells it at $\$\frac{3}{4}$, does he gain or lose? and how much?

6. Find the sum of $\frac{4}{5}$, $7\frac{1}{2}$, and $8\frac{3}{4}$.

7. $\frac{5}{17} + \frac{3}{34} + \frac{16}{136} + \frac{1}{2} = ?$

8. A man engaged to labor 30 days, but was absent $5\frac{7}{12}$ days; how many days did he work?

10. A young man received a salary of $\$60\frac{2}{3}$ a month, and paid for his board $\$30\frac{1}{2}$, for washing $\$1\frac{1}{4}$, and for other expenses $\$12\frac{9}{10}$. How many dollars had he left?

384. 1. Define a proper fraction, and give an example of one.

2. A merchant bought three pieces of cloth containing $125\frac{1}{2}$, $96\frac{3}{4}$, and $48\frac{2}{3}$ yards. How many yards did he buy?

3. What is the value of $2\frac{1}{2}$ times $\frac{3}{4}$ of $\frac{4}{5}$ of $1\frac{1}{3}$?

4. If 9 men consume $\frac{3}{4}$ of $9\frac{3}{4}$ pounds of meat in a day, how much does one man consume?

5. A farmer distributed 15 bushels of corn among several persons, giving them $1\frac{2}{3}$ bushels apiece. Among how many persons did he divide it?

6. What is the value of $\dfrac{11\frac{3}{4}}{\frac{4}{7}}$?

7. What number must be added to $22\frac{2}{3}$ that the sum may be $99\frac{1}{3}$?

8. A can do a piece of work in 8 days, and B can do it in 6 days. In what time can they do it working together?

9. A pole stands $\frac{1}{4}$ in the mud, $\frac{1}{4}$ in the water, and 21 feet above the water. What is its length?

10. A man bequeathed to his son $3500, which was $\frac{5}{7}$ of what he left his wife. How much did he leave his wife?

385. 1. If $\frac{3}{8}$ of a farm is valued at $1728, what is the value of the whole?

2. If 8 be added to both terms of the fraction $\frac{5}{8}$, will its value be increased or diminished? and how much?

3. If the sum of two fractions is $\frac{5}{8}$, and one of them is $\frac{9}{20}$, what is the other?

4. Express in its simplest form the quotient of 2025 divided by 3645.

5. If the dividend is $\frac{7}{8}$ and the quotient $\frac{4}{39}$, what is the divisor?

6. At $\$\frac{1}{4}$ a bushel, how many bushels of apples can be bought for $\$5\frac{1}{8}$? (Analysis.)

7. Define fraction, terms of a fraction, improper fraction, compound fraction, and complex fraction.

8. Change $\frac{524}{16}$ to a whole or mixed number.

9. How many 8ths of a bushel in $9\frac{1}{2}$ bushels?

10. Change $\frac{2}{3}$, $\frac{1}{4}$, $\frac{7}{15}$, $\frac{5}{12}$, $\frac{4}{8}$ to equivalent fractions having a common denominator.

386. 1. A farmer sells 6 jars of butter holding 8 pounds, at 36¢ a pound, and receives in payment 14 cans of coffee, each holding two pounds. What was the price of the coffee? Work by using cancellation.

2. If a man walks $3\frac{4}{5}$ miles in one hour, how far can he walk in 9 hours?

3. Find the sum of $13\frac{2}{3}$, $1\frac{1}{4}$, $6\frac{5}{8}$, $20\frac{19}{24}$, and $\frac{7}{18}$.

4. How many days' work at $\$1\frac{3}{4}$ a day will pay for $8\frac{13}{16}$ yards of cloth at $\$2\frac{1}{2}$ a yard, and 56 lb. of butter at 25 cents a pound?

5. The product of two numbers is $41\frac{2}{3}$, and one of them is $160\frac{9}{144}$; what is the other?

6. Find the sum of $93567 + 20754867 + 4756 + 925674 + 6543987 + 6579 + 98675 + 567923 + 645876 + 9346 + 878 + 54562 + 888.$

7. Tell in words what these numbers are: 1950; 90; 4040; 73000007.

8. Find the difference between 76392 × 4506 and 985301 × 976.

9. What will 79 ten-ton cars of coal be worth at $5.50 a ton?

10. If you should buy 376 horses for $65123, how much would you sell them for apiece to gain $5189?

DECIMALS.

387. 1. A merchant bought four pieces of cloth containing $32\frac{5}{8}$, $38\frac{1}{2}$, $40\frac{7}{8}$, $45\frac{3}{4}$ yards, respectively. How many yards did he buy?

2. Change to decimals and add: $\frac{7}{8}$, $\frac{3}{4}$, $4\frac{3}{8}$.

3. From a farm containing $128\frac{7}{8}$ acres, $84\frac{5}{8}$ acres were sold. How many were left?

4. Find the prime factors of 1008.

5. $50 \div .05 = ?$

6. What will 2.47 pounds of coffee cost at $.48 per pound?

7. If one yard of ribbon costs $34\frac{1}{2}$ cents, what will 6 pieces cost, each piece containing 13.12 yards?

8. Write in words 68.0642.

9. If 9 yards of cloth cost $1.17, what will 15 yards cost?

10. A lady went shopping with $45. She paid 4\frac{1}{10}$ for shoes, 5\frac{2}{3}$ for a hat, 12\frac{3}{4}$ for a dress. How much money had she left?

388. 1. If a farm is worth $3200, how much is $\frac{2}{8}$ of it worth?

2. From one million take one millionth.

3. What is the difference in cents between $\frac{3}{4}$ of a dollar and $\frac{1}{3}$ of a dollar?

4. Point off into periods 96308796, and write over each period its name.

5. Express with figures the following numbers: Seven million ninety-five thousand, sixty-three and fifteen thousandths, and seven hundred and seven hundredths.

6. Read (write in words) the following: 642.0016; 100.01.

7. 353812416 ÷ 589 = ?

8. Find the sum of 684.8, 96.84, 6.075, .1906, 7508.

9. At $9 per M., what will 6728 feet of lumber cost?

10. At $.65 per C., what will 1240 pens cost?

389. 1. Reduce to a simple fraction $\dfrac{\frac{5}{8} \text{ of } \frac{3}{7}}{\frac{6}{11} \text{ of } 1\frac{1}{3}}$.

2. What fraction of $11\frac{3}{4}$ is $5\frac{2}{3}$?

3. What common fraction equals .0125.

4. Reduce $\$\frac{3}{16}$ to a decimal.

5. A farmer sold 120 sheep, which were $\frac{5}{8}$ of his flock. How many had he before the sale?

6. A grocer sold $\frac{1}{3}$ of a barrel of sugar to one man and $\frac{1}{4}$ of it to another, and had 80 pounds left. How many pounds did the barrel contain at first?

7. A and B can do a piece of work in 12 days, A can do it in 25 days; in how many days can B do it?

8. A man spent $\frac{2}{5}$ of his money for a horse and $\frac{3}{4}$ of the remainder for a buggy and harness, and had $37.50 left. How much money had he at first?

9. In dividing by a decimal, how do you determine the proper place of the decimal point in the quotient?

10. At $9.75 per thousand, what will 16544 bricks cost?

390. 1. Which is the greater, $\frac{19}{21}$ or $\frac{23}{35}$? How much greater?

2. Find the sum of $78\frac{3}{4}$, $87\frac{3}{10}$, $4\frac{5}{8}$, and $79\frac{1}{2}$.

3. A man has three lots, which are 120, 420, and 600 ft. wide respectively. He wishes to divide them into lots of the greatest equal width possible. How wide will each lot be? How many such lots can he make?

4. If .375 of a ton of coal cost $2.40, what is the price of one ton? How many tons can be bought for $80?

5. Reduce to decimals $\frac{3}{4}$, $\frac{1}{8}$, $\frac{5}{8}$, $\frac{17}{20}$, $\frac{1}{18}$.

6. Write in figures thirteen thousandths, four hundred and five hundredths, five hundred fifteen millionths, and add the results.

7. Reduce to a simple fraction $\dfrac{\frac{5}{16} \text{ of } \frac{4}{15}}{\frac{3}{7} + \frac{5}{8}}$.

8. Sold a house for $4,797, which was two-sevenths more than it cost; find the cost price.

9. Make a bill for the following articles, bought to-day of James Brown, No. 23 Warburton Avenue, Yonkers, N. Y.: 30 oranges at 25 cents a dozen; 7 lb. of coffee at 28¢; $3\frac{1}{2}$ lb. prunes at 13¢; 1 bag of sugar containing 28 lb. at $5\frac{1}{2}$¢. Receipt the bill as though you were James Brown's clerk.

10. Bought three boxes of oranges containing 263, 220, and 156, at $3.50 per hundred, and sold them for 50¢ per doz. Find the amount of profit.

391. 1. $\dfrac{(7\frac{1}{8} - 2.05) \div (5 \times .23)}{.7\frac{2}{3} + 2.23\frac{1}{3} - .6 \div .4}$.

2. $1000 \div .001 = ?$

3. $\frac{1}{2} + \frac{1}{4} + .75 + 1\frac{2}{3} + .330 = ?$

4. Reduce $\frac{7}{18}$ to a decimal.

5. $\frac{7}{10} + \frac{8}{100} - \frac{13}{1000} + \frac{5}{20} + \frac{3}{18} = ?$

6. $\frac{1}{2} \times 10.0019 \times 1.2 \times \frac{3}{8} \times .463 = ?$

7. Change .0507 to hundredths.

8. Change 8.84 to a common fraction in its lowest terms.

9. $.123 - .01 - .11 - .003 = ?$

10. $.0509 + \frac{7}{8} - .27 = ?$

392. 1. $(.05015 \div 2.006) + \overline{(24.6 \div .0012} \times \frac{1}{18}) - 1200\frac{5}{8} = ?$

2. Express in words the following: 10020.00042024; .000702; .00000018; 30000.00030; .00010020.

3. If a man travels at the rate of 7.4 miles an hour, how long will it require to travel 370 miles?

4. What will be the cost of $3\frac{1}{2}$ yd. of cloth at .75 dollars per yard?

5. Find the sum of .125, 46.42, 9.3, 164.25, .80406.

6. From 1000 subtract .001.

7. Find the cost of 445.375 bushels of wheat at $.9173 per bushel.

8. Change .00125 to a common fraction.

9. Reduce $\frac{4}{18}$ to a decimal.

10. At $.044 per pound, how many pounds of sugar can be bought for $44?

393. 1. Find the cost of $9\frac{1}{2}$ tons of coal, if $\frac{3}{4}$ of a ton cost $3.00.

2. Find the sum of 40 units, 20 tens, 464 thousandths, 5 ten-thousandths, and 1 millionth.

3. Write in figures two and twenty-six hundredths; two and twenty six-hundredths.

4. What number multiplied by $14\frac{1}{4}$ will produce $1684\frac{1}{2}$?

5. If $\frac{4}{7}$ of a yacht is valued at $\$3840\frac{1}{2}$, what is the value of the whole?

6. If $\frac{5}{8}$ of a pound of tea cost $\$.50$, what will $16\frac{3}{4}$ pounds cost?

7. Reduce to simplest form:

$$\tfrac{3}{4} \text{ of } \frac{4\frac{5}{9}}{6\frac{1}{8}} \times \frac{6\frac{8}{11}}{11\frac{4}{7}}$$

8. Reduce $\frac{113}{125}$ to a decimal.

9. A man bequeathed $\frac{7}{13}$ of his estate to his elder son, and the remainder to his younger son, who received $\$1344$. What was the estate worth?

10. What must be paid for 8960 pounds of plaster at $\$5.50$ per ton?

DENOMINATE NUMBERS.

394. 1. Define simple quantity; compound quantity.

2. Reduce 1760 cwt. to higher denominations.

3. Add: 11 oz. 11 pwt. 15 gr.; 7 oz. 12 pwt. 19 gr.; 10 oz. 13 pwt. 17 gr.

4. Write the table of long measure.

5. Find the total area in sq. yards of the ceiling of a room 18 ft. long, and 15 ft. wide.

6. Find the number of square feet in the surface of a cube 3 ft. by 3 ft. by 3 ft.

7. Find the total area in the four walls of a room 18 ft. long, 15 ft. wide, and 9 ft. high.

8. Define fraction; mixed number; proper fractions; improper fractions.

9. Reduce $\frac{2}{3}$, $\frac{3}{8}$, and $\frac{5}{24}$ to similar fractions.

10. Define circumference; diameter.

395. 1. What is the value of $\frac{2}{3}$ of $\frac{3}{4}$ divided by $\frac{1}{2}$ of $\frac{2}{3}$ plus $\frac{3}{4}$ of $\frac{4}{3}$?

2. A and B can build a shop, working together, in 10 days; B can build it, working alone, in 30. In how many days can A build it?

3. Add 0.525 mi., 0.125 rd., 0.5 yd., and 0.16 ft.

4. From $\frac{2}{11}$ of a square rod take $\frac{3}{4}$ of a square yard.

5. Find $\frac{5}{7}$ of 9 A. 70 sq. rd. 15 sq. yd. 7 sq. ft. 19 sq. in.

6. There is a room 15 ft. long, 12 ft. wide, and 9 ft. high; it has 2 windows, each 3 ft. by 6 ft., and a door 3 ft. by 7 ft. Taking out the space for door and windows, how much will it cost to plaster this room at 25¢ per square yard?

And what will be the cost of floor boards $1\frac{1}{4}$ in. thick, to lay the floor of this room at $40 per thousand?

7. There is a square field 40 chains around; how many acres are in it?

8. In a space 27 ft. long, 18 ft. wide, and 12 ft. high, there may be placed how many cubes 3 feet on each edge?

9. How many grains in a ton? How many gallons in a cu. yard?

10. How many grains in a Troy pound?

396. 1. What decimal of a mile is $\frac{1}{7}$ of 5 mi. 89 rd. 3 yd. 2 ft.?

2. Divide 15 T. 17 cwt. 29 lb. 7 oz. by $\frac{4}{5}$.

3. $36\frac{1}{5}$ sq. in. equals what fraction of an acre?

4. $\frac{1}{8}$ mi. + $\frac{2}{3}$ rd. + $\frac{1}{2}$ ft. − $7\frac{1}{2}$ yd. = ?

5 If 7 spoons weigh 7 oz. 12 pwt. 9 gr., what will 13 similar spoons weigh?

6. Add $36\frac{3}{8}$, .00125, 1460, $\frac{3}{4}$, $\frac{5}{16}$, and 16.26.

7. If $\frac{3}{8}$ of a ship is worth $6285, what is $\frac{5}{18}$ worth?

8. To-day you, as a clerk of Chester & Wilson, sell Wm. Lambert 20 bbl. flour at $4.87½, 4500 lb. meal at $1.06 per cwt., and 2450 lb. bran at $13.50 per T. Make out the proper bill.

9. Define improper fraction; decimals; reduction descending; a bill.

10. Reduce £17 14s. 3far. to farthings, and prove.

397. 1. For 22 lb. 14 oz. of butter worth 16¢ a pound a man gets 12 quarts of sirup. What is the price of the sirup per gallon?

2. At $3 a perch, what would masons earn in laying a wall 8 ft. high and 2 ft. thick in a cellar dug 36 ft. × 42 ft.?

3. At 60¢ per yd., what will be the least cost to carpet a room 14 ft. × 16 ft. with ingrain carpet, using only full breadths, and no waste for cutting?

4. Reduce .875 of a bushel to lower denominations.

5. How many bushels will a bin contain that is 9 ft. long, 4 ft. wide, and 6 ft. deep?

6. How much will a piece of land 20 rd. by 18 rd. cost at $116 per A.?

7. Find the cost of a Brussels carpet (27 in. wide) at $1.15 per yd. for a room 16 ft. by 23 ft., breadths to run crosswise.

8. At $.60 per sq. yard, what will it cost to plaster sides and ceiling of a room 18 × 12 × 8 ft.?

9. Leaving Lockport, I travel until my watch is 1 h. 20 min. slow. Which way, and how far, have I travelled?

10. My cistern is 8 ft. by 4½ ft. When the water is 27 in. deep, how many barrels of water is there in the cistern?

TEST QUESTIONS.

398. 1. Define and illustrate decimal; multiple; quotient.

2. If I burn a pint of kerosene every night, what will a three weeks' supply cost me at 15 cents a gallon?

3. Find the sum of $\tfrac{7}{8}$ mi. $\tfrac{1}{3}$ rd. $\tfrac{5}{8}$ ft.

4. How many boards, each 15 feet long, will be required to build $56\tfrac{4}{11}$ rods of fence four boards high? Analyze.

5. Find the value of $\tfrac{3}{8}$ of a chest of tea weighing $57\tfrac{1}{2}$ pounds, at $\$1\tfrac{1}{2}$ per pound.

6. Solve $\dfrac{14 \times 32 \times 96 \times 7 \times 163}{192 \times 21 \times 28 \times 55 \times 8} = ?$

7. How many times will a wheel 12 ft. 4 in. in circumference revolve in going 10 miles?

8. How many days must a laborer work, at $\$1.12\tfrac{1}{2}$ a day, to pay for 6 cords of wood, at $\$3.37\tfrac{1}{2}$ per cord?

9. A man was born Feb. 29, 1844, and died Mar. 15, 1880. How many birthdays did he have? What was his age?

10. What is the product of 12 millionths multiplied by 12 thousandths?

399. 1. How many pickets 3 in. wide, placed 3 in. apart, will be required for a fence around a rectangular yard 4 rd. 6 ft. long, and 3 rd. 8 ft. wide?

2. A farmer has a piece of land containing $7\tfrac{13}{8}$ acres, fenced in the form of a rectangle, its length being twice its width. What are the dimensions of rectangle?

3. Oswego County has an area of 970 square miles, and a population of 71780. What is the population to the square mile? How many acres could be given to each one of the entire population?

4. Oswego is in longitude 76° 35′ W., Albany, 73° 32′ W. What is the difference in their longitude? When it is noon in Albany, what o'clock is it in Oswego?

5. What will be the cost of carpeting a room 18 ft. long and 12 ft. wide with Brussels carpet ¾ yd. wide, at 85¢ a yd., the strips to run lengthwise of the room, and allowing 4 in. to be turned under?

6. At $25 per thousand, what is the value of 16 planks, each 18 ft. long, 6 in. wide, 2½ in. thick?

7. Find the cost of 5 pieces of timber, each 48 ft. long, 9 in. by 12 in., at $1.50 per hundred bd. ft.

8. How many board feet of lumber will be required to fence a lot 80 ft. by 40, the boards being 10 ft. by 6 in., and the fence 4 boards high?

9. How many board feet will it take to cover the top of a tank 14 ft. long, 6 ft. wide, with boards 2 in. thick?

10. A man sold two bushels of strawberries as follows: to Mrs. A. he sold $\frac{1}{12}$ of the berries, to Mrs. B. ⅜, and the remainder to Mrs. C. How many quarts did Mrs. C. buy?

400. 1. Two telegraph stations are 18 miles, 224 rods apart. If the telegraph poles between the stations are 8 rods apart, how many poles will be needed, and how much will they cost at 50¢ apiece?

2. What is the value of a triangular piece of land, having a base of 60 chains and an altitude of 40 chains, at $60 per acre?

3. How many times can a dish holding 2 qt. ½ pt. be filled from a jar holding 3 gal. 2 qt. 1 pt.? How much will be left in the jar?

4. Find cost of carpeting a room 24 ft. long and 18 ft. wide, with carpet 27 inches wide, the strips running lengthwise of the room, cost of carpet being $1.65 a yard, and no loss in matching the figures.

5. After spending $46⅔, I had ⅜ of my money left. How much had I at first?

6. A man traded 7 wagons at $\$77\frac{1}{2}$ each for 84 bbl. of flour; what was the flour per barrel?

7. What is the capacity in liters of a cistern 1.5 meters long, 9 decimeters wide, and 86 centimeters deep?

8. How many bricks 8 in. long, 4 in. wide, and 2 in. thick will it take to pave a section of street 200 ft. long, 36 ft. wide, the bricks being placed on their edges?

How much will the bricks cost at $7.35 per M.?

9. What is the depth of a cubical bin which contains 300 bu. of wheat?

10. The distance around a circular park is $1\frac{1}{2}$ miles. How many acres does it contain?

401. 1. How many blocks $\frac{11}{16}$ of a foot long can be cut from a board 22 ft. long?

2. How many poor families can be supplied with $\frac{2}{7}$ of a ton of coal each from 12 tons?

3. How many pairs of tray-cloths, each containing $\frac{3}{4}$ of a yard, can be cut from 15 yards of linen?

4. In how many months, paying $\$\frac{3}{4}$ per week, will a debt of $36 be paid?

5. $\frac{1}{3}$ is what part of $\frac{3}{4}$?

6. A 37-gallon cask is $\frac{5}{6}$ full; $6\frac{1}{2}$ gallons being drawn off, how full will it be?

7. If from a piece of cloth containing 96 yd. you sell $24\frac{2}{3}$ yd., what fractional part of the piece remains?

8. $11\frac{3}{4}$ bushels are what fraction of $15\frac{2}{3}$ bushels?

9. $\frac{3\frac{1}{2}}{6}$ is what part of $\frac{2}{3}$ of $\frac{7}{8}$?

10. A man had 700 head of cattle. He sold at one time 50 head, at another 75 head. What fraction of the whole did he sell?

402. **1.** How many cubic feet of stone will it take to build the walls of a cellar 36 ft. long, 24 ft. wide, and 8 ft. high, outside measurement, the walls being 18 in. thick ? How much will the stone cost at $4.50 per cord ?

2. Find the diameter of a wheel whose circumference is 50 feet.

3. If 1 bu. 3 pk. 6 qt. of walnuts cost $3.10, what is the price per quart?

4. What will be the cost of 5 gal. 3 qt. 1½ pt. of maple sirup at 75 cents per gallon?

5. Find the cost of 5362 pounds of coal at $4.50 per ton.

6. How long a time has elapsed since the first message was sent by telegraph, May 29, 1844 ?

7. How much profit will there be in buying 4 bu. 1 pk. 6 qt. of cranberries at $2 a bushel, and selling them at 10 cents a quart?

8. How many days will a 6-ounce bottle of medicine last a patient who takes a teaspoonful three times a day, a teaspoon holding 60 drops or minims ?

9. Multiply 9 mi. 25 rd. 3 yd. 2 ft. by ⅜.

10. Divide 110 mi. 149 rd. 3 yd. 2 ft. 6 in. by ⁵⁄₇.

403. **1.** Multiply 25 yards 2 ft. 11 in. by 16.

2. From 6 bu. 6 qt. take 3 pk. 1 qt. 1 pt.

3. What is the difference in time between June 16, 1890, and Feb. 4, 1895?

4. What will it cost to build the walls of a cellar that is 26 ft. long and 16 ft. wide, 6½ ft. deep, the wall being 18 in. thick, at $1.50 a perch ?

TEST QUESTIONS. 255

5. A field is 16 ch. 10 links long and 5 ch. wide. How many acres does it contain?

6. How many board feet in 24 joists, 10 in. by 2 in. by 16 ft., and what are they worth at $11 per M.?

7. What is a pile of four-foot wood worth that is 16 ft. long and 6 ft. high, at $4.50 a cord?

8. How many grains in 5 lb. of butter?

9. Reduce 12 cwt. 80 lb. 6 oz. to the decimal of a ton.

10. Find the sum of 184⅜, 372½, 19⅝.

PERCENTAGE.

404. 1. Express as % the following: .28; .065; 3.07; .004.

2. Express decimally the following: ½%; 6¼%; 8%; 125%.

3. From a farm of 144 acres 18 acres were sold. What per cent of the farm was sold?

4. A grocer sold eggs at 12½ cents a dozen and gained 25%. What was the cost?

5. A man's farm cost him $5,400; his crop of potatoes yielded him in cash 8% of the cost of the farm. What was the value of his potatoes?

6. If a merchant pays $.80 a yard for a roll of carpet, and because it became damaged sells it for $.65 a yard, what per cent does he lose?

7. Sent my agent in St. Louis $3017.60, with which he is to purchase flour at $4.00 per bbl., after deducting his commission at 2½ per cent. How many barrels should I receive?

8. If, by selling 36840 ft. of lumber at $21.12 per M., you gain 28 per cent, what would be your gain or loss by selling it at $17 per M.?

9. If a merchant has marked an article for sale at 50 per cent above cost, what per cent will he deduct from the asking price if he sells the article at cost?

10. $7884.00 is to be raised by taxation in a certain school district. The taxable property of the district is $584,000. Find the rate of tax, and A's tax, whose property is assessed at $3850.

405. 1. From $\frac{1}{3}$ of a week take $\frac{1}{3}$ of a day.

2. Reduce $\frac{11\frac{2}{3}}{12\frac{2}{3}}$ to a simple fraction.

3. Define base and rate.

4. How many hundredths of anything is $\frac{1}{2}$ of it? $\frac{1}{4}$ of it? $\frac{1}{3}$ of it? $\frac{1}{10}$ of it?

5. What is 12% of 1682?

6. Express as common fractions in their lowest terms: 25%, 62½%, 12½%, 16⅔%.

7. A speculator bought 2160 barrels of apples, and upon opening them found 15% of them spoiled. How many barrels did he lose?

8. A farmer sold 50 sheep, which was 25% of his whole flock. How many sheep had he at first?

9. My income this year is $4028, which is 24% less than it was last year. How much was it last year?

10. A commission merchant sells goods to the amount of $6895. What is his commission at 3%?

406. 1. I bought two houses at $3500 each, and sold one at a gain of 22%, and the other at a loss of 22%. Did I gain or lose on both? and how much?

2. If I sell for $16 what cost $20, what per cent do I lose?

3. If I buy a piano for $450, and sell it for $600, what per cent do I gain?

4. Define insurance; premium; taxes.

5. What will be the cost of insuring a quantity of wheat valued at $8,450, at ⅜%?

6. The premium for insuring a schoolhouse, at the rate of 1¼%, was $75. For what sum was it insured?

7. The town of B is to be taxed $3,700 to build a bridge; the taxable property is valued at $1,850,000. What will be the rate of taxation, and the tax of Mr. A., whose property is valued at $5,000?

8. What is the duty, at 25%, on 4796 pounds of Russia iron, worth 10 cents a pound?

9. What number increased by 25% of itself is 506.25?

10. Find the net cost of a bill of goods amounting to $3,750 at 10% discount, and 4% off for cash.

407. 1. An agent sold 4,250 yd. of calico at 3⅞¢ per yard. What was his commission at 2¼%?

2. A real estate broker, who charges 4% commission, receives $224 for selling a house. What price is paid for the house?

3. If $8,240 is sent to an agent to cover the amount of his purchase and his commission of 3%, what is the amount of his purchase?

4. A hotel is insured for $90,000 at 2⅛% for 3 years. What is the annual cost of insurance?

5. A man's weight is 180 pounds, and he is 20% heavier than his brother. What is his brother's weight?

6. A bill for hardware amounting in gross to $2,537.75 is subject to discounts of 40%, 10%, and 5%. What is the net amount?

7. If you remove the decimal point from the number 6.45, what effect does it produce upon the number?

8. If from the same number you take the period from after the 6 and place it before the 6, what will be the effect?

9. At $12.75 a ton what will 3265 pounds of hay cost?

10. A tree measures 8.2 ft. in circumference. What is the diameter?

408. 1. Find $\frac{1}{4}\%$ of $12.00; $\frac{2}{10}\%$ of 2000 bushels of corn; 200% of 5 dozen eggs; $\frac{1}{3}$ of 1 per cent of 100 tons of coal.

2. What fraction increased by 25 per cent of itself equals $1\frac{1}{6}$?

3. What is the effect upon the quotient when both the dividend and the divisor are multiplied by the same number?

4. Express as fractions in lowest terms, $8\frac{1}{3}\%$, $2\frac{1}{2}\%$, $18\frac{3}{4}\%$.

5. Express as per cent, using the sign, .1352, $\frac{1}{4}$, 2, $\frac{1}{300}$.

6. Express as decimals, $\frac{1}{400}$, $\frac{1}{3}$, $\frac{1}{4}\%$, 20%, $15\frac{1}{2}\%$.

7. What per cent of the number of days in February, 1896, is the number of days in January, 1896?

8. My house cost $6000, which was 400 per cent more than I paid for the lot. Find the cost of both.

9. After spending $14 for a suit of clothes, a man had $126 left. What per cent of his money did he spend?

10. An agent purchased $8\frac{1}{2}$ tons of sugar at $3\frac{1}{2}$ cents per pound on 3% commission. Find the cost of the sugar, including commission.

409. 1. What is the rate of taxation on $1000 when $147000 is raised on $35,000,000?

2. A man selling cloth at $4.20 per yard, gained 20%. Had he sold it at $3.60 per yard, would he have gained or lost? and what per cent?

3. If $\frac{5}{8}$ of a mill is worth $10000, what is $\frac{1}{2}$ of the remainder worth?

4. Bought a horse for 160\frac{1}{2}$, and sold it for $\frac{5}{8}$ of its cost. How much did I lose?

5. Define least common multiple; improper fraction; prime factor.

6. Simplify $\dfrac{9\frac{2}{3}}{2\frac{1}{4}}$.

7. Find the cost of 10 sticks of timber, each 16 feet long, 14 inches wide, and 10 inches thick, at $16.50 per M., board measure.

8. How many gallons will a cistern hold that is 12 ft. long, 8 ft. wide, and 6 ft. deep?

9. If $9\frac{3}{4}$ yards of cloth are worth $24.375, what is the value of $16\frac{7}{8}$ yards at the same rate?

10. Name the unit of weight in the metric system, and give the table in which that unit occurs.

410. 1. I spend 65 per cent of my salary, but am able to save $980; how much do I spend?

2. How much must I send my agent that he may buy, at $1\frac{1}{2}$ per cent commission, 400 bbl. flour at $6.75 per bbl.?

3. Given the amount and percentage, write the formula for finding each of the other terms.

4. What are like numbers? Unlike numbers?

5. Write an abstract number. Give definition of abstract number.

6. Write in words, 2300406.000960.

7. What kind of number is 4.6 bushels?

8. A father divided his property as follows: to his son John he gave $\frac{1}{4}$, to his daughter Susan $\frac{1}{5}$, to his wife $\frac{1}{3}$, and the rest, which was $13000, to endow a school. What was the value of his estate?

9. I own a house that cost me $3000. It cost me to insure it for 3 years $24. The average yearly cost of repairs is $50. The average yearly tax is 2% of the cost. I can get 5% per annum for the $3000 invested. The house will last 60 years. I receive in rent for the house $300 per annum. If these conditions are constant, how much will I gain or lose in 60 years?

10. A father is 39 years old and his daughter 13; what per cent of the father's age is the daughter's?

411. 1. Write these per cents as hundredths: 2%, $6\frac{1}{2}$%, 20%, $12\frac{1}{2}$%.

2. How many per cent of a number is 0.20? 0.75? .12$\frac{1}{2}$? 1.40?

3. What fractions of a number (in lowest terms) are these per cents: $16\frac{2}{3}$%? 75%? $33\frac{1}{3}$%? 100%? and 175%?

4. Express as hundredths and as common fractions: $\frac{1}{4}$%; $\frac{3}{8}$%; $\frac{1}{2}$%; $\frac{3}{8}$%; and $\frac{1}{10}$%.

5. From a stack of hay 7 T. 11 cwt. were sold, which was $75\frac{1}{2}$% of the whole. How much did the stack contain before the sale?

6. A lawyer collected 65% of a debt of $1260, and charged 5% commission on the sum collected. What did the creditor receive?

7. If a hat that cost $5 be sold for $9, what is the gain per cent?

8. How many days from Sept. 16, 1892, to Feb. 12, 1894?

9. 874 is $33\frac{1}{3}\%$ less than what number?

10. Required the cu. feet of a box 6 ft. 6 in. by 4 ft. 9 in. by 3 ft. 3 in.

412. 1. Write the following numbers and add: six thousand sixteen and sixty-five thousandths, four hundred one thousand forty-one and one-tenth, six hundred one and nine hundredths, ten thousand one hundred seventeen and nine hundred three thousandths, forty-nine hundred forty-nine and nine-tenths.

2. Write in words $83.493, 7007\frac{7}{10}, 1001001.01, 90019\frac{4}{100}$.

3. Find the number of which 160 is $\frac{2}{3}$.

4. Find the exact number of days from July 4, 1893, to to-day.

5. Multiply 7 lb. 8 oz. 15 pwt. by 15.

6. Solve $\dfrac{18 \times 963 \times 44 \times 27 \times 2800}{63 \times 88 \times 105 \times 1926 \times 45} = ?$

7. Define commission, also brokerage; and state on what sum, or value, both are computed.

8. Express decimally $27\frac{2}{3}$ and $\frac{3}{16}$. Find their product as decimals, and as common fractions, expressing both answers decimally.

9. Fruit was sold at $12\frac{1}{2}¢$ per quart, which was 200 per cent of its cost. What was the cost per bushel? and what was the rate per cent of profit?

10. An agent sold 840 bu. grain at 60¢ per bushel. His commission was $15.12. Find the rate of commission.

413. 1. A man owes you a debt of $2160, which he declines to pay. Your lawyer succeeds in collecting 70 per cent of the debt, and charges 5 per cent commission for his services. What sum do you receive?

2. A manufacturer sent $1295.27 to a commission merchant who charges 3 per cent commission, instructing him to purchase wool at $0.33⅓ per pound. How many pounds of wool will be received?

3. A farm was sold for $8000, which was 20 per cent less than its real value. If it had sold at $12000, what per cent above its real value would it have brought?

4. A commission merchant sold for a farmer 6000 lb. pork at 8½¢ per pound. He charged 1½% commission for selling, and paid $18.80 for freight. How many feet of pine boards at $25 per 1000 ft. could he purchase with the proceeds of the pork, after deducting 1 per cent commission for buying?

5. Reduce to simple fraction in lowest terms:

$$\frac{\frac{2}{11} \text{ of } 12\frac{2}{7}}{\frac{3}{8} \times \frac{5}{6} + \frac{2}{3}}.$$

6. What per cent of 3 is ⅔? Of ⅘ is ⅔? Of 80 is 50?

7. A drover sold 250 sheep for $1150, which was 15% more than they cost. What was the cost per head of the sheep?

8. If 20% be lost on a ton of rye straw sold for $19.20, what is the cost of the straw per ton?

9. How many per cent of a number is 0.15? 0.06¼? 0.50? 2.25?

10. What common fraction of a number in its lowest terms is 20%? 50%? 6½%? 66⅔%? 160%?

414. 1. A man sold $8400 worth of merchandise, and had 30% of his stock left. What was his entire stock worth?

2. A merchant sold goods at 20% and 5% off, and still made 20% on the cost. What was the cost price of a book that was marked $1.00?

3. Bought 1000 pounds of butter at 18¢, and sent it to an agent who sold it at 21¢ on a 5% commission. What was my rate of gain?

4. Mr. Brown has a flock of 940 sheep in three fields. In the first are 20% of the entire flock, in the second 40%, and the remainder in the third. How many sheep are there in each field?

5. A lady has a salary of $825 a year; she spends 20% of it for board, 35% of it for other expenses, and saves the remainder. What sum does she save?

6. What per cent of a leap year is the time from Washington's Birthday to the Fourth of July?

7. The Barber Asphalt Company engaged to pave a street 5 miles long at $55000 a mile. If the actual cost be $130 per rod, what is the gain per cent?

8. A commission merchant charges $1\frac{1}{2}$% for selling, and $2\frac{1}{4}$% for guaranteeing the payment of the money. His commission on a certain transaction amounted to $384.75. Required the amount of the sale.

9. I bought 1100 tons of coal at $3\frac{1}{2}$ per ton. I sold 40% of it at a gain of 50%, 40% of the remainder at a gain of 35%, and lost 10% on the rest. What was my actual gain?

10. An article bought at 18% below the asking price is sold for the asking price. What is the gain per cent?

INTEREST AND DISCOUNT.

415. 1. Find the amount of $875 for 1 year, 4 months, and 12 days, at 6 per cent interest.

2. Find the interest on $128.45 from March 2, 1895, to Dec. 14, 1895, at 6 per cent.

3. A pile of wood 256 feet long, 4 feet wide, and 5 feet high is sold for $160. What is the price per cord?

4. Define per cent; interest; proper fraction.

5. State the difference between a prime and a composite number.

6. Find the cost of 6 gal. 3 qt. and 1 pt. of sirup at 46 cents per gallon.

7. 1521 is how many times 13?

8. What is the interest on $1200 for 2 yr. 3 mo. 18 da. at 6%? The amount?

9. What is the interest on $1240 from March 3 to Aug. 28, at 6%?

10. Write the United States rule for computing the amount due on a note when partial payments have been made.

416. 1. In what time will $3960 earn $770 at 5%, simple interest?

2. If $675, at simple interest, gain $172.80 in 3 years, 2 months, 12 days, what is the rate of interest?

3. When interest, time, and rate are given, how may the principal be found?

4. Define true present worth and true discount of a debt. Define compound interest, and make and solve an example to illustrate your definition.

5. A merchant sells goods amounting to $6784.00 on a year's credit. If money is worth 8%, what sum should he accept in payment of the bill 6 months before it becomes due?

6. Write a negotiable promissory signed by James Fox for $875.60 due 90 days from date, payable to yourself, at a bank. Name (*a*) the payee; (*b*) the drawer; (*c*) the date when the note matures (becomes due). What words on the note make it negotiable? What does negotiable mean?

7. If you should sell the note (Ex. 6) to Mr. F. P. Weaver, what indorsement must you write upon it? Where should indorsements be written?

8. If the note is not paid until Sept. 15, 1895, how much interest will then be due on it?

9. A farmer expended $5580 in improvements on his farm, which was 24% more than ⅜ of the cost of the farm. Find the cost of the farm.

10. Principal, interest, and time being given, how is the rate found?

417. 1. Find the amount of $496.85 for 2 years, 4 months, and 15 days at 4 per cent.

2. How long will it take $750 at 6 per cent to gain $67.50 interest?

3. A dealer bought 65 lawn-mowers at $4.25 each, and sold them at $3.87½ each. What per cent did he lose?

4. If a cellar is 38 ft. long and 28 ft. wide inside the wall, and the wall is 8 ft. high and 18 in. thick, how many cubic yards of masonry does the wall contain?

5. What per cent of a number equals ⅝ of the number? What part of a number equals 33⅓ per cent of it?

6. Write decimally, 6%; one hundred six per cent.

7. A town 6 miles long and 4½ miles wide is equal to how many farms of 80 acres each?

8. What number must be subtracted from four hundred sixty-seven thousand six hundred thirty-three to make it exactly divisible by 758?

9. Find the amount of $535.20 for 2 yr. 4 mo. 18 da. at 5 per cent, simple interest.

10. Give formula or rule for finding the base when rate per cent and difference are given. Form and write such a problem.

418. 1. Find the interest of $263.75 for 1 yr. 3 mo. 16 da. at 5%.

2. Make a 30-day bank note dated Jan. 20, 1896, for $600, payable at some bank. Find date of maturity, the discount, and proceeds if discounted on the date of the note. (Make the note on a separate piece of paper, and have it properly indorsed.)

3. What is the present worth of a note due in 1 yr. 6 mo. ?

4. In what time will $600 gain $30 interest at 6% ?

5. What will $300 amount to in 4 years compounded annually at 4% ?

6. An agent says he will insure your house for 3 years at 65. What does he mean by "at 65"?

7. Define interest; principal; usury; compound interest.

8. Find the amount of $684.50 for 3 yr. 4 mo. at 7%.

9. Compute the interest of $1250 for 2 yr. 5 mo. 18 da. by the six per cent method.

10. What is the interest on a note for $515.62, dated March 1, 1885, and payable July 16, 1888 ?

419. 1. A note for $710.50, with interest after 3 mo. at 8%, was given Jan. 1, 1884, and paid Aug. 13, 1886. What was the amount due ?

2. What sum of money will gain $173.97 in 4 yr. 4 mo. at 6% ?

3. What is the legal rate of interest in this State?

4. Find the exact interest of $950 at 5% for 98 days.

5. What principal will amount to $1531.50 in 1 yr. 3 mo. 6 da. at 6% ?

6. At what rate will $1500 amount to $1684.50 in 2 years, 18 days ?

TEST QUESTIONS. 267

7. In what time will $840 gain $78.12 at 6%?

8. How long will it take any sum of money to double itself at 4%?

9. Find the compound interest of $460 for 1 yr. 5 mo. 24 da. at 6% interest, payable semi-annually.

10. If ¾ of an acre of land costs $15, what will 10½ acres cost?

420. 1. Name four different forms of reduction of common fractions. Illustrate one of them to show that the value of the fraction remains unchanged.

2. Define simple interest, true discount, and bank discount. How does bank discount differ from interest? How does it differ from true discount?

3. Define cancellation, and state the principle of arithmetic that authorizes its use.

4. Find the amount of $575.87½ at 5 per cent, simple interest, from Aug. 5, 1883, to March 17, 1885.

5. What principal will earn $71.68 in 2 years, 4 months, at 6 per cent, simple interest.

6. At what rate, simple interest, will $175 amount to $203.35 in 3 yr. 7 mo. 6 days?

7. In what time will $4260 earn $873.30, at 6 per cent?

8. A 60-day note for $610.25, dated June 12, 1889, was discounted in bank, July 1, at 6 per cent. Find the term of discount, discount, and proceeds.

9. Having purchased a horse for $125, you wish to borrow that amount at bank for 6 mo. Write your own note, indorsed by your parent as security, for the sum which, discounted to-day, will give $125 as proceeds of the note.

10. A stock of goods was owned by three parties. A owned ⅜, B ⅜, and C the remainder. The goods were sold at a profit of $4260. What was each one's share of the gain?

421. 1. A horse is offered me for $350 cash, or for $382.50 to be paid in 4 mo. What can I save by paying cash, the rate of interest being 6%?

2. Which is the more profitable, and how much, money being worth 5%, to buy a house for $5940 on 2 years' credit, or for $5219.30 on 6 months' credit?

3. A note dated June 20, 1893, and bearing interest at 6 per cent, was paid Aug. 15, 1895. The face of the note being $68.45, what was the amount paid?

4. Bought 150 front feet of land at $40 per front foot, paid $116 city taxes, $32 county taxes, and $320 local taxes; at the end of two years I sold for $60 per front foot. Reckoning interest at 6% on the purchase price, did I gain or lose by the transaction? and how much?

5. A man wishes to pay me $3252.56. Not having the money, he borrows it from a bank by giving his note for 48 days at 4%. For what sum does he draw the note? No grace.

7.

$545.50.   Buffalo, N. Y., *Apr. 2, 1896.*

Sixty days after date, I promise to pay ~~~~~Henry Hamilton~~~~~ or order, Five hundred forty-five and 50/100 Dollars. Value received.

                            Chas. C. Trowbridge

This note was discounted May 4, 1896. Find the proceeds.

7. Required the simple interest and amount of $7231.289 for 3 yr. 8 mo. 15 days at 8%.

8. Face of note $750. Time 60 da. Rate 6%. To find proceeds.

9. Write the following in a note properly, and find the maturity and proceeds: Face, $600; date, April 3, 1896; due in 90 days; discounted at bank, May 20, 1896, at 6%, with grace.

10.

$9000.   Saratoga Springs, N.Y., Oct. 3, 1895.

Nine months after date, I promise to pay to the order of ~~~~~~~~~Gates & Co.~~~~~~~~~ Nine thousand Dollars, at the First National Bank. Value received.   S. B. Graves.

Find proceeds, if discounted at 6%, Dec. 3, 1895.

STOCKS AND AVERAGE PAYMENTS.

422. 1. How many shares of stock at 80 can I buy for $2550?

2. I sold two houses for $2400 each. On one I gained 10%, on the other I lost 10%. How much did both cost me? Did I gain or lose in the whole trade? and how much?

3. Find the cost of 40 shares of American Express Co. stock at 105½, brokerage ½%.

4. A mining company declares a dividend of 8% per annum on its stock. What is the nominal value of a man's shares who gets $864 as his semi-annual dividend?

5. If the stock of a railway company sells at 5% above par, what will 25 shares cost?

6. If I invest $21,008 in 5% bonds at 104, what will be my annual income?

7. Sugar bought at 5 cents a pound was sold for 6½ cents; what per cent was gained?

8. What sum invested in 4 per cent stock will yield an annual income of $320, if you purchase stock at par?

9. What would be your investment, if the stock is worth 15 per cent above par?

10. A man invested his money in 6% railroad stocks, and received $300 semi-annually. What was the sum invested?

423. 1. What sum must be invested in stocks bearing 6½ per cent interest, at 105 per cent, to produce an annual income of $1000? Solve by cancellation.

2. Define brokerage, certificate of stock, par value, premium (as used in stocks and investments). What are bonds? Name some of the different classes of bonds.

3. What income will be realized from investing $4190.63 in 5% stock, purchased at 7% discount, if I pay ⅛% for brokerage?

4. What is the value of 31 shares of $500 each, sold at a premium of $2\frac{5}{100}$%?

5. Which is more profitable, to buy 8% bonds at 25% premium, or 6% bonds at 10% discount?

6. A owes B $3000, due as follows: June 15, $1500; Sept. 10, $400; Nov. 1, $500; Dec. 15, $600. B accepts in settlement Oct. 26 a note for 9 months, bearing interest at 6% for the amount of the debt, with 6% interest due him at that date. Find face of note.

7. On Jan. 1, 1895, a merchant gave three notes: one for $500, payable in 30 days; one for $400, payable in 60 days; and one for $600, payable in 90 days. What is the average term of credit, and what the equated time of payment?

8. E. R. Smith owes J. D. Wilson $2500, due Oct. 12, 1896. If Mr. Smith pays $500 Aug. 10, and $1000 Sept. 25, when should the balance be paid?

9. A speculator bought N. Y. C. stock at $98\frac{1}{2}$, and sold it at $97\frac{3}{4}$, and lost $187.50. How many shares did he handle?

10. Had he retained his stock until a quarterly dividend was declared, his dividend would have been $312.50. What was the annual rate of dividend?

424. 1. State why securities fluctuate in value.

2. Name a corporation.

3. What does a stockholder hold to show that he has stock in a company?

4. On what does the income from his stock depend?

5. Why does a corporation issue bonds?

6. Find the present worth and true discount of $300, due in 10 months, at 6%.

7. Find the bank discount and proceeds of a note of $730, due in 3 months, at 6%.

8. What is the face of a note at 2 months and 18 days, which yields $2961 when discounted at a New York bank?

9. A person owning $\frac{3}{8}$ of a piece of property, sold 20% of his share. What part did he then own?

10. At what price should $4\frac{1}{2}$% bonds be bought to make the income from investment equivalent to that from 3% bonds at par?

PROPORTION AND PARTNERSHIP

425. 1. What is ratio?

2. Read the following: 3 : 15. What does it equal?

3. What is each of the numbers in the above expression called?

4. What is a proportion?

5. Is the following expression a proportion? Explain why. 9 : 12 :: 16 : 24.

6. 24 : ( ) = 56 : 7. Find the omitted term.

7. If 8 men can do a piece of work in 10 days, in how many days can 12 men do it?

8. If 3 men in 12 days of 10 hours each can build a wall 100 feet long, 14 feet high, and 3 ft. thick, how long will it take 4 men working 8 hours a day to build a wall 200 feet long, 16 feet high, and 4 feet thick?

9. If it takes 5 men 4 hr. 24 min. to manufacture 400 boxes, how much time will 8 men require to perform the same work?

10. If $\frac{2}{3}$ of an acre of land cost $15, what will $10\frac{1}{2}$ acres cost?

426. 1. 50 men in 7 da. at 12 hours a day dig a cellar. How many men will be required to dig a similar cellar in $21\frac{3}{5}$ da. of 8 hr. each?

2. A and B enter into partnership, A with $1800 and B with $900. After 8 mo. B adds $300 to his capital. Divide a profit of $840 between them at the end of the year.

3. A bankrupt owes A $350, B $680.50, C $65, D $500, E $980.50; his property nets $1648.64. How much does each creditor receive? How much does he pay on a dollar?

4. What is the ratio of 7 to 8? Of $2\frac{1}{2}$ to $3\frac{1}{3}$? Of $9 to $6?

5. If 20 men can mow a field in 6 days, in how many days will 30 men mow it?

6. If 5 horses eat 8 bu. 14 qt. of oats in 9 days, at the same rate how long will 66 bu. 30 qt. last 17 horses?

7. A and B hired a pasture for $40 for the season. A put in 9 cows for 4 mo., and B put in 8 cows for 8 mo. Other conditions being the same, what should each pay?

8. In what time will $10,000 yield $1200 interest at 8%. Solve by proportion.

9. If the antecedent is $\frac{2}{3}$ of $\frac{9}{12}$ of $\frac{6}{20}$, and the ratio is $\frac{5}{8}$ of $\frac{1\frac{2}{3}}{}$ of $\frac{1\frac{2}{3}}{}$, what is the consequent?

10. Required the ratio of $6\frac{1}{2}$ cu. ft. to $11\frac{2}{3}$ cu. ft.

427. 1. A, B, and C entered into partnership. A put in $600 for 8 mo., B $800 for 7 mo., C $1500 for 4 mo. They gained $820. What was each one's share of the gain?

2. A, B, and C found a gold-mine, and after developing it sold it for $64000. They agreed to divide the money according to the time each had worked. A worked 37 days, B 46 days, and C 39 days; for extra services B is to receive $1800, and C $1200 additional. How much does each receive?

3. Three men, A, B, and C. enter into partnership. Out of a gain of $1200, C takes $500 and B $400. A's investment is $4500. Find B's and C's investment.

4. Divide $450 among three people in the ratio of 3, 4, and 8.

5. Three persons bought a block for $21000, of which A paid $9000, B $8000, and C the remainder. They rented it for $1400 a year. What was each man's share of the rent?

6. Forster, Stull, and Furlong made 8000 pairs of bicycle pedals in 1895, which they sold for $1.60 per pair. The pedals cost them $1.15 per pair. If Mr. Forster put in $1000 Jan. 1, Mr. Stull $1200 April 1, and Mr. Furlong $900 May 1, what would be each one's share of the gain after drawing out the original investment?

274 SENIOR ARITHMETIC.

7. Four men purchased a city block for $36,000. The first contributed $20,000, the second $7,000, the third $4,000, and the fourth $5,000. They sold the land at an advance of 50% on the purchase price. How much was each man's share of the gain?

8. A, B, and C form a partnership in which A is to furnish no capital, but give his whole time to the business, and have ½ the profits. B furnishes $10,000, and C $15,000. Their net profit at the end of a year is $8000. What is each partner's share?

9. A, B, and C gain in business together respectively $700, $1000, and $1500. What was the investment of each if their joint capital was $16,000?

10. Smith, Brown, and Jones gain in trade $9400. Smith furnished $10,000 for 5 months, Brown $9000 for 6 months, Jones $7000 for 1 year. Apportion the gain.

INVOLUTION AND EVOLUTION.

428. 1. Define involution; evolution; a square; cube root.

2. Find the square of $6\frac{2}{3}$; of 2.35.

3. Find the third power of 123.

4. Find the square root of the fraction $\frac{2601}{3025}$.

5. What is the distance around a square field which contains 40 acres?

6. A man has 640 acres of land. How much more will it cost to enclose it with a fence at $4 a rod, in a rectangular form 512 rods long and 200 rods wide, than it would if in the form of a square?

7. What is the length of one side of a cube which contains 8120601 cubic inches?

8. Find the entire surface of a cube whose volume is 42 cu. ft. 1512 cu. in.

9. The edge of a cube is 42 inches. Find the length of the edge of another cube 4 times as large.

10. If 16 cords of wood be piled in the form of a cube, what will be the length of one of its edges?

429. 1. What are the length and breadth of a rectangular field which contains 60 acres, the length of which is three times its breadth?

2. A rectangular farm of 300 A. is $7\frac{1}{2}$ times as long as it is wide. How many miles of fence will enclose it?

3. A bird is 15 feet above a monument 80 ft. high. A boy is 145 ft. from the bird. How far is the boy from the base of the monument?

4. How far is it between the extreme corners of a box 10 ft. square and 6 ft. deep?

5. Find how many acres in a lot in the form of a right-angled triangle whose hypothenuse is 50 rd. and the base 40 rd.

6. Find the diagonal of a square piece of land equal in area to a rectangular piece whose dimensions are 80 rd. by 20 rd.

7. Wishing to know the height of a church steeple, I find it casts a shadow 165 ft.; I also find that a 10-ft. pole when placed perpendicular casts a shadow $12\frac{1}{2}$ ft. What is the height of the steeple?

8. A house is 36 ft. wide, and the ridge of the roof is 12 ft. above the plates. How long are the rafters?

9. A steamer goes due north at the rate of 12 miles an hour, and another goes due east at the rate of 15 miles an hour. How far apart will they be at the end of 8 hours?

10. If a pineapple 5 in. in diameter costs 20¢, what should be the cost of a pineapple of similar shape 6 in. in diameter?

MISCELLANEOUS.

430. 1. The sum of two numbers is 2120, and their difference 938. What is each number?

2. J. & R. Ross, New York, bought of A. L. Covert & Co., Philadelphia, the following articles, June 20, 1881: 15 Nichols's Geography at $0.65; 12 Meiklejohn's Literature at $0.80; 25 Bowser's Geometry at $0.75; 15 Hawthorne & Lemmon's Literature at $1.12; 10 Thomas's U. S. History at $1.00.

They paid $25 in cash, and returned books to the amount of $10. Make out bill showing entire statement.

3. Oswego, N.Y., contains 22,000 inhabitants. If each inhabitant should contribute one cent per week for fifty-two weeks towards the erection of a soldiers' monument, how expensive a monument could be built at the end of the year?

4. The State of New York has 7746 miles of railroad, which cost $588,672,762. Find the average cost per mile.

5. The sum of three numbers is 96: the least is $4\frac{1}{2}$, and greatest $37\frac{2}{3}$. Find the other number and the product of the three numbers.

6. $9,000,000 has recently been appropriated for improving the Erie Canal. If it is 352 miles long, how many dollars may be expended on each mile?

7. Find the least common multiple of 24, 60, 75, 120.

8. What is the smallest sum of money with which I can purchase oxen at $30 each, cows at $60 each, or horses at $80 each?

9. Find the difference between the G. C. D. and the L. C. M. of 81, 45, 108, and 135.

10. What is the greatest number that will exactly divide 3640, 12750, and 18755?

MISCELLANEOUS.

431. 1. If the ties on the N.Y.C. & H.R.R. are $1\frac{2}{3}$ ft. apart from centre to centre, how many are there from New York to Buffalo, a distance of 450 miles?

2. If the Empire State express has an average rate of 62 miles an hour, how many hours and minutes will it take to run from Syracuse to Albany, a distance of 150 miles?

3. Multiply $7\frac{2}{3}$ by $17\frac{1}{3}$.

4. E. C. Stearns & Co. sell 24 bicycles at $\$62\frac{1}{2}$ apiece; what do they bring?

5. How many times does a bicycle wheel $9\frac{2}{3}$ ft. in circumference revolve in going 3 miles, there being 5280 ft. in a mile?

6. Multiply $\dfrac{\frac{1}{2}-\frac{1}{6}}{\frac{1}{2}+\frac{1}{3}} + \frac{1}{2}$ by $12\frac{1}{2}$.

7. A and B can build a house in 30 days: B can do the work alone in 45 days. In how many days can A do it alone?

8. Write a complex fraction, whose numerator shall be a simple fraction, and its denominator compound.

9. A drover bought 375 sheep at $\$4\frac{1}{2}$ per head. He sold 200 of them at a loss of 20 cents per head, and gained enough on the rest to balance the loss. What did he receive per head for the rest?

10. A can do a piece of work in 5 days; B can do the same work in 8 days. In what time can they do it working together?

432. 1. A boy paid for a book $.70, which was $\frac{5}{8}$ of his money. The remainder he spent for marbles at $2\frac{1}{2}$ cents apiece. How much money had he at first? and how many marbles did he buy?

2. At a school examination $\frac{5}{7}$ of the pupils passed, and 250 pupils failed. How many pupils were examined? and how many passed?

3. $\frac{1}{2}$ of a number diminished by $\frac{1}{3}$ of it is equal to 5. What is the number?

4. $\frac{4}{21}$ of 1743 is $\frac{83}{112}$ of what number?

5. A man after giving $\frac{1}{3}$, $\frac{1}{4}$, and $\frac{1}{5}$ of his money in charity had $10000 left. How much had he at first?

6. Four persons own a ship. A owns $\frac{1}{4}$ of it, B $\frac{1}{3}$ of the remainder, C $\frac{1}{2}$ of what then remained, and D the remainder, which is worth $3000. What is the value of the ship?

7. If $\frac{2}{3}$ of a number be divided by 4, and $\frac{1}{3}$ of $\frac{1}{4}$ of the number be taken from the quotient, the remainder will be 6. What is $\frac{2}{3}$ of the number?

8. One person can do a piece of work in 6 days, another can work twice as fast. How long will it take them to do the work together?

9. A boy was asked how many fish he had caught. He said that the difference between $\frac{1}{2}$ and $\frac{2}{3}$ the number was six. How many had he?

10. A, B, and C can do a piece of work in 5 da. A can do it alone in 12 da., C can do it in 15 da.; in what time can B do it?

433. 1. What will it cost at $1.75 a yard to carpet a floor 18 ft. long, 14 ft. wide, with carpet $\frac{3}{4}$ yd. wide?

2. How many yards of carpeting 27 inches wide will be required for a room 30 ft. long, 24 ft. wide, if the strips run crosswise, and 6 inches be allowed for matching?

3. What fraction of a great gross is 3 gross, 5 doz., $1\frac{5}{7}$ units?

4. At $.27 per sq. yard, find the cost of plastering a room 30 ft. by 24 ft. by 12 ft. high, allowing for a baseboard one foot high, two doors 9 ft. by 3 ft., and 5 windows 6 ft. by 3 ft.

5. Reduce 5 cd. ft. 9$\frac{3}{8}$ cu. ft. to the fraction of a cord.

6. Reduce 33 gal. 3 qt. 1 pt. 1$\frac{7}{13}$ gi. to the fraction of a hhd.

7. How much tin will be required to make a pail and cover, the pail to be 6 inches in depth and 7 inches in diameter, and the rim of the cover to be 1 inch deep?

8. At $16.50 per M., what will be the cost of 12 sticks of timber, each 14 ft. long, 10 in. wide, and 8 in. thick?

9. How many board feet in a plank 16 ft. long, 15 in. wide at one end, and 10 in. wide at the other end, and 3 in. thick?

10. The longitude of New York is 74° 0' 3" W., and that of San Francisco 122° 23' W. When it is 1 P.M. at New York, what is the time at San Francisco?

434. 1. The longitude of Syracuse, N.Y., is 76° 9' 16" W., and that of Berlin, Germany, is 13° 23' 44" E. When it is noon in Berlin, what is the time at Syracuse?

2. The Oswego River is 24 miles long, and descends 120 feet in that distance. What is the average descent per mile?

3. Add $\frac{2}{3}$ A., $\frac{1}{4}$ sq. rd., $\frac{1}{4}$ sq. yd., $\frac{3}{4}$ sq. ft.

4. Find the cost of 4 T. 7 cwt. 40 lb. of hay at $12 per ton.

5. From a cask containing 44 gal. 2 qt. 1 pt. of vinegar, 8 gal. 3 qt. leaked out. What decimal of the original contents remained?

6. Find the number of square inches in the surface of a block 2 ft. long, 18 in. wide, and 10 in. high.

7. The sun rose in the latitude of New York, April 1, 1896, at 5 o'clock and 43 minutes, and set at 6 o'clock and 25 minutes. It rose April 30 at 4 o'clock and 59 minutes, and set at 6 o'clock and 55 minutes. How much longer was the thirtieth day than the first?

8. How long and wide must a granary be to hold 4000 bushels of grain, if it is 8 ft. high, and the grain to be placed in bins 6 ft. back on each side of an aisle 4 feet wide?

9. A cubic foot of water weighs $62\frac{1}{2}$ pounds. How many barrels in a cistern of water that weighs 6 T. 5 cwt.?

10. Find the cost of 1 bu. 1 pk. 1 qt. and 1 pt. of chestnuts at 5¢ per quart.

435. 1. At what rate per cent will $2500 gain $625 in 3 years, 4 months?

2. A merchant buys goods at $1.20 a yard, and, after keeping them 6 mo., sells them at $1.35. What is his rate of gain?

3. A man buys oranges at 1¢ each, and sells them at 18 cents a dozen. What is his gain per cent?

4. Find the amount on $836.22 from Feb. 19, 1895, to June 3, 1896, at 6%.

5. 3200 votes are cast for two men; one has a majority of 374. How many votes did each receive?

6. A man borrowed $756.12, June 28, 1872. What must he pay to cancel the debt July 11, 1872, at 6%?

7. A commission merchant in Minneapolis received $6150, with directions to purchase flour. His terms were $2\frac{1}{2}$% on the amount purchased. How many barrels of flour at $3 a barrel can he ship to the sender of the money?

8. A merchant sells goods at an advance of 20%, but loses 5% of his sales by bad debts. What % does he gain?

9. A bought a carriage at 20% discount with 10% and 5% off, and sold it at the list price. What % profit did he make?

10. An agent sold some Western land, and paid to the former owner $7531.30, retaining $153.70 as commission. What rate did he charge?

436. 1. A district schoolhouse cost $8010; the valuation of the property of the district is $392,375, and the number of polls assessed at $1.25 each is 130. What is the rate of tax, and what was A's tax, who paid for 4 polls, the valuation of his property being $6000?

2. What sum of money placed on interest at 6% will amount to $1567.85 in 1 year, 3 months?

3. Sold wheat at 72 cents per bushel, and thereby lost 10% of the cost. What was the cost per bushel?

4. What will be the net cost of stationery billed at $850, if the discount is 20% and 10% off?

5. A house worth $7200 is insured for ⅝ of its value, at the rate of 60 cents on $100. Find the premium.

6. A man sold a house for $4200, which was 20% more than it cost him. What did it cost?

7. On a bill of goods listed at $645, choice is given between discounts of 20%, 10%, and 5% off, or a direct discount of 35% off. Which is better? and how much?

8. If a merchant gains $16\tfrac{2}{3}\%$ by selling cloth at $1.40 per yard, find his gain on a sale amounting to $32.

9. I owe B a bill of $1980. If I borrow the money from a bank, what must be the face of a note, due in 60 days without interest, which I must give to the bank, that I may receive the amount necessary to pay him, discount at 6%?

10. A man sells his house for $8000, and receives in payment a note for 90 days. After 30 days he has the note discounted at a bank at 6%. What does he receive for it?

437. 1. I was offered $160 cash for my buggy, or a note of $165 payable in 90 days. I took the note, and discounted it at a bank at 5%. Did I gain or lose? and how much?

2. What is the difference between the true and bank discount on $1250 for 90 days at 6%?

3. If John lends James $300 for 4 months, how long ought James to lend John $800 to equal the favor?

4. I have a note of $1225, due in 48 days. Needing the money immediately, I get it discounted at a bank at 6%. How much shall I receive? and how much will the bank take? No grace.

5. Three men hire a pasture for $60. A put in 4 cows for 11 weeks, B 5 cows for 12 weeks, and C 8 cows for 5 weeks. What ought each to pay?

6. If a man 5 ft. 10 in. high casts a shadow 4 ft. 6 in. long, what is the height of a tree which casts a shadow 85 feet long at the same time?

7. Give the inverse ratio of $\frac{3\frac{3}{4}}{4\frac{2}{5}}$ to $\frac{6\frac{2}{3}}{8\frac{1}{3}}$.

8. Required the ratio of £21 15s. to £6 15s.

9. A, B, and C entered business with a certain capital, Jan. 1, 1894. Jan. 1, 1896, they find the business to be worth $7000, which is a gain of 40% on the original capital. A's share of the gain is 50%, B's share 30%, and C's share 20%. What amount did each invest?

10. What did each gain?

MISCELLANEOUS.

438. 1. If 4 barrels of flour will last 3 persons for 1 year, how many barrels will be required to last 10 persons 10 months?

2. The shadow of a flag-staff at a certain time of day was 64 feet in length. A line stretched from the top of the flag-staff to the extremity of the shadow measured 150 feet. Required the height of the staff.

3. Messrs. Stevens, Jones, & Payne form a partnership, placing into their business $350, $450, $1500 respectively. They make $570 the first year. What share of the profits should each receive?

4. By selling 3% stock at par, and buying 4% at 110, a man increases his income $105 a year. How many shares of the 3% stock does he sell?

5. A, B, and C entered into a partnership. A furnished $1200 for 8 mo., B furnished $1600 for 9 mo., and C furnished $1000 for a year. They lose $560. What is each man's loss?

6. What is the length of a walk laid diagonally through a park which measures 60 rods on one street and 80 rods on another?

7. What will be the difference in ratio of income between 5% stock bought at 120 and 4% bought at 95?

8. A father dying left to his family a certain sum of money, of which the wife received $8000, his daughter $4000, and each of two sons $6000. What part of the whole did each receive?

9. If sugar costs 5½ cents per pound and coffee 33 cents per pound, what is the ratio of the cost of the sugar to that of the coffee?

10. At $.50 per rod, how much will it cost to enclose a field of 80 acres, that is twice as long as it is wide?

SENIOR ARITHMETIC.

439. 1. If the sale of coal at $.75 per ton above cost yields a profit of $18\frac{3}{4}\%$, how much must the seller add to this price to make a profit of 40% ?

2. At what price must a 4% stock be purchased to yield 5% on the investment ?

3. If a pile of wood 38 ft. long, 4 ft. wide, and 5 ft. high costs $6250, what will be the cost of a pile 64 ft. long, 8 ft. wide, and 6 ft. high ?

4. If 10 men, working 10 hr. a day for 13 da., can build a fence 200 rd. long, how many men, working 11 hr. a day for 10 da., can build 92 rd. of the same kind of fence ?

5. The smaller of two numbers is 36, and one half of the ratio between it and the larger is 2. What is the larger number ?

6. What number has the same ratio to 5 that $\frac{1}{4}$ has to $\frac{1}{8}$?

7. Find the mean proportional between 16 and 36. Between $\frac{1}{10}$ and 1.

8. What income on his investment will a man realize if he purchases 4% stock at 125 ?

9. If A's capital is $3000, and B's $2000, how much more should B invest at the end of 6 mo. that he may share equally with A at the end of the year ?

10. What is the rate per cent of a tax for 52.88\frac{1}{4}$ on property assessed at $3525.50 ?

440. 1. Write your own promissory note for $200, with interest payable in ninety days from to-day to any person you choose.

2. On what month and day would your note become due, including days of grace ? Give one reason why the note is void (worthless). Find the amount due on your note at its maturity.

MISCELLANEOUS. 285

3. What is the time of day when the time past noon equals the time to midnight? When $\frac{1}{2}$ the time past noon equals the time to midnight? When the time past noon equals $\frac{1}{4}$ the time to midnight?

4. A cask can be emptied by a $\frac{1}{2}$-inch faucet in 4 hours. In what time can it be emptied by a $1\frac{1}{2}$-inch faucet?

5. Explain the difference between factor and root; between product and power.

6. A and B divide $90 in the ratio of $\frac{2}{3}$ to $\frac{5}{6}$. What is each one's share?

7. If a tank $13\frac{1}{2}$ ft. long, $7\frac{1}{4}$ ft. wide, and $3\frac{1}{3}$ ft. deep holds $73\frac{1}{3}$ barrels of water, how wide must another tank be that is 9 ft. 9 in. long, 4 ft. 10 in. deep, and holds $89\frac{1}{2}$ barrels?

8. $\dfrac{1 + (\frac{2}{3})^3 - \sqrt[3]{\frac{1}{27}}}{7 \times (\frac{1}{3})^3} = ?$

9. A milkman's quart measure is too small by one gill. At 5 cents a quart, how much does he dishonestly make in the month of June, if he sells 500 false quarts daily?

10. Find the length of the diagonal of an are of land in the form of a square.

MENSURATION.

441. The process of measuring lines, surfaces, and solids is **Mensuration**.

442. A **Line** is that which has length, without breadth and thickness.

443. A **Straight Line** is the shortest distance between two points, or a line that does not change its direction at any point.

444. A **Curved Line** changes its direction at every point.

445. A Plane Surface is a surface that does not change its direction.

446. A **Quadrilateral** is a plain figure having four straight sides.

447. Parallel Lines are lines having the same direction and equally distant from each other.

448. A **Parallelogram** is a quadrilateral whose opposite sides are parallel.

449. A **Rhomboid** is a parallelogram whose angles are not right angles.

What is a parallelogram called whose angles are right angles?

450. A **Rhombus** is a rhomboid whose sides are equal.

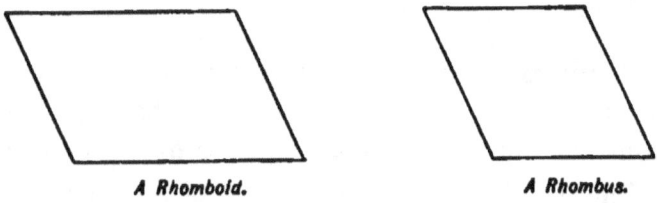

A Rhomboid.   *A Rhombus.*

451. The area of a rhomboid is found by multiplying the base by the altitude.

NOTE. — The altitude of a parallelogram is the perpendicular distance between the sides.

1. Find the area of a parallelogram whose base is 24 rods, and altitude 18 rods.

2. Find the area of a rhombus whose base is 15 ft. and altitude 8 ft.

3. Draw a rhomboid whose base is 15 ft. and altitude 10 ft. Find its area.

Draw a rectangle having the same dimensions.

MENSURATION. 287

452. A **Trapezoid** is a quadrilateral having only two sides parallel.

453. **To find the area of a trapezoid,** multiply ½ the sum of the parallel sides by the altitude.

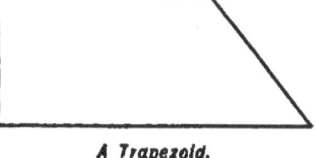
A Trapezoid.

4. Find the area of a trapezoid whose altitude is 10 ft., its longest side 20 ft., and shortest side 15 ft.?

5. A board 20 inches wide at one end and 12 inches wide at the other is 16 feet long. How many board feet does it contain?

454. A **Trapezium** is a quadrilateral having no two sides parallel.

NOTE. — By drawing a diagonal between any two opposite sides of a trapezium, we have two triangles, the diagonal serving as the base of each. The altitude of each is the perpendicular distance from its other angle to the diagonal.

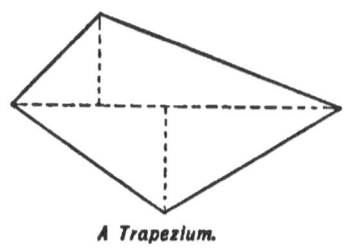
A Trapezium.

455. **To find the area of a trapezium,** multiply the diagonal by half the sum of the altitudes of the two triangles.

6. The diagonal of a trapezium is 18 ft.; the altitudes of its two triangles are 5 ft. and 3 ft. What is the area?

7. A farm is in the form of a trapezium. The diagonal distance between the northern and southern corners is 108 rods, and the perpendicular distances from the east and west corners to the diagonal are 52 rods and 36 rods respectively. How many acres in the farm?

SOLIDS.

456. A solid whose two bases are equal and parallel, and its other faces parallelograms, is called a **Prism.**

NOTE. — Prisms take their names from the form of their bases, as triangular, quadrangular, pentagonal, hexagonal, etc., according as the bases have three, four, five, or six sides, etc.

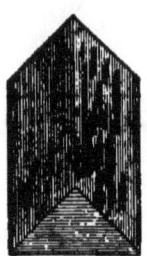
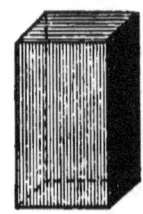

A Triangular Prism.   *A Rectangular Prism.*

457. To find the contents of a **prism,** multiply the area of the base by the altitude.

8. Find the contents of a triangular prism whose altitude is 10 in., and the area of its base 7 sq. in.

9. What are the contents of a quadrangular prism whose base is 5 in. by 8 in., and whose altitude is 12 in.?

10. What are the contents of a hexagonal prism, the area of whose base is 10 sq. ft., and whose altitude is 15 ft.?

PYRAMIDS AND CONES.

458. A solid whose base is a triangle, square, pentagon, etc., and whose sides are triangles meeting at a vertex, is called a **Pyramid.**

NOTE. — A pyramid takes its name from the form of its base.

A solid whose base is a circle, and whose convex surface terminates in a point, is called a **Cone.**

459. The **Altitude** of a pyramid or cone is the perpendicular distance from its vertex to the centre of its base.

The **Slant Height** is the shortest distance from the vertex to the perimeter of the base.

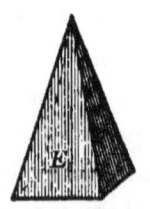
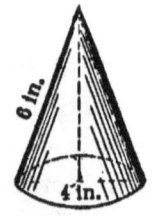

A Pyramid.   *A Cone.*

460. To find the contents of a pyramid or cone, multiply the area of the base by ⅓ of the altitude.

To find the convex surface, multiply the perimeter of the base by ½ the slant height.

11. Find the contents of a quadrangular pyramid whose altitude is 10 in., and whose sides of bases are 8 in. and 6 in.

SOLUTION. — 8 × 6 × $\tfrac{10}{3}$ = 160 cu in. *Ans.*

12. Find the convex surface of a regular hexagonal pyramid whose slant height is 16 in., and whose side of base is 4 in.

13. Find the convex surface of a cone, when the circumference of its base equals 16 ft. and its slant height 18 ft.

14. Find the convex surface and volume of a cone whose radius is 4 in. and altitude 6 in.

461. The **Frustum** of a cone or pyramid is the part which is left after the top is cut off in a plane parallel to the base.

462. To find the contents of the frustum of a pyramid or cone, multiply ⅓ of the altitude by the sum of the areas of the two bases plus the square root of their product.

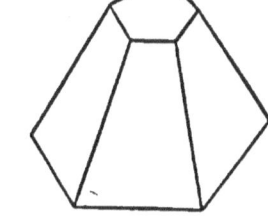 

Frustum of a Pyramid.   *Frustum of a Cone.*

15. Find the contents of the frustum of a quadrangular pyramid whose altitude is 15 ft., and whose ends are 6 ft. and 4 ft. square.

16. A log 16 ft. long is 30 in. in diameter at one end and 24 in. at the other. Find its cubical contents.

463. A **Sphere** is a solid bounded by a curved surface, all parts of which are equally distant from the centre.

464. To find the surface of a sphere, multiply the circumference by the diameter.

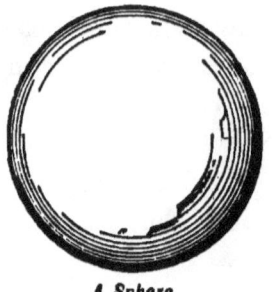

A Sphere.

465. To find the contents of a sphere, multiply the surface by ⅙ of the diameter.

17. Find the surface of a sphere when the diameter is 16 inches.

18. Find the surface of a sphere when the radius is 3 yards.

19. Find the surface of a sphere when the radius is 5 cm.

Find the volume when:

20. Diameter = 25 ft.   22. Radius = 12 ft.
21. Radius = 2 ft.   23. Circumference = 125664 in.
24. Radius = 3 dm.

25. Compare the volume of a 4-ft. cube and a 4-ft. sphere.

26. Compare the surfaces of a 4-ft. cube and a 4-ft. sphere.

REVIEW OF MENSURATION.

466. 1. Find the cubic yards in a cone, the circumference of whose base is 20 ft., and whose altitude is 30 ft.

2. Find area of a semi-circle when its radius equals 14 ft.

3. Find area of a square inscribed in a circle of 4 ft. in diameter.

4. The circumference of a circle and the perimeter of a square are each 300 ft. Which has the greater area?

5. A circle is inscribed in a 6-ft. square. Find the area of the circle.

MENSURATION. 291

6. Find the value at $50 an acre of a farm in the form of a trapezoid, the parallel sides of which are 120 rd. and 160 rd. respectively, the distance between which is 80 rd.

7. How many miles does the earth travel in a revolution around the sun, the distance between them being 95,000,000 miles?

8. If a bin is 8 feet square, how deep must it be to hold 100 bushels?

9. Find the lateral surface of an equilateral triangular pyramid, the perimeter of the base being 12 m. and the slant height 14 m.

10. Find the volume of a square pyramid, the perimeter of whose base is 16 feet, and whose altitude is 9 ft.

11. What is the volume of the largest cone that can be cut from a pyramid whose base is 6 feet square, and whose slant height is 15 feet?

12. A cylindrical tank is 14 ft. deep and 6 feet in diameter. Find the cost of cementing sides at 90¢ a sq. yard.

13. Find the capacity in gallons of a cylindrical cistern whose inside diameter is 6 feet, and whose depth is 7 feet.

14. Find the capacity in Kl. of a cylindrical cistern whose inside diameter is 4 m., and whose altitude is 5 m.

www.ingramcontent.com/pod-product-compliance
Lightning Source LLC
Chambersburg PA
CBHW032048230426
43672CB00009B/1517